How the Ray Gun Got Its Zap

How the Ray Gun Got Its Zap

ODD EXCURSIONS INTO OPTICS

Stephen R. Wilk

OXFORD
UNIVERSITY PRESS

Oxford University Press is a department of the University of Oxford.
It furthers the University's objective of excellence in research, scholarship,
and education by publishing worldwide.

Oxford New York
Auckland Cape Town Dar es Salaam Hong Kong Karachi
Kuala Lumpur Madrid Melbourne Mexico City Nairobi
New Delhi Shanghai Taipei Toronto

With offices in
Argentina Austria Brazil Chile Czech Republic France Greece
Guatemala Hungary Italy Japan Poland Portugal Singapore
South Korea Switzerland Thailand Turkey Ukraine Vietnam

Published in the United States of America by
Oxford University Press
198 Madison Avenue, New York, NY 10016

Library of Congress Cataloging-in-Publication Data
Wilk, Stephen R.
How the ray gun got its zap : odd excursions into optics / Stephen R. Wilk.
pages cm
Includes bibliographical references. ISBN 978-0-19-994801-7 (acid-free paper)
1. Optics. 2. Optics—History. 3. Optics—Biography. 4. Optics—Social aspects—
History. I. Title.
QC361.W55 2013
621.36—dc23
2013009902

1 3 5 7 9 8 6 4 2
Printed in the United States of America
on acid-free paper

{CONTENTS}

PART II Weird Science

PART III Pop Culture

How the Ray Gun Got Its Zap

Introduction

The essays in this book began as a result of my taking over the editorship of the *Light Touch* column in *Optics and Photonics News*, a news and information journal (as opposed to an academic journal) published by the Optical Society of America, in 2003. I had already written one *Light Touch* piece for them ten years earlier, "How Retroreflectors *Really* Work," which appeared in the December 1993 issue (and which appears, greatly rewritten and expanded, as chapter 25 in the current collection). I wanted to steer the column away from the subjects and styles that it had been covering—optics demonstrations and experiments suitable for pre-college students—and toward what that early article of mine had covered—optics esoterica and brief discussions, accompanied by some humor. That, to me, was what the series title suggested—Education by Stealth, as the BBC so succinctly put it. This would also provide an excuse and opportunity to explore many odd facets of optics that I'd encountered over the years. My first outing was "Yes, I was right, it is obvious!," which was successful in that it generated a surprising amount of mail, about evenly divided between condemning my suggestion and praising my bringing to light a little-appreciated side of optics. The folks at *OPN* were very pleased. Although I was to get a lot of feedback over the past decade, nothing has approached the attention that initial article produced.

Not long after, I started doing brief pieces for *The Spectrograph*, the house organ of the George Harrison Spectroscopy Laboratory at MIT. I was acting as an unpaid Visiting Scientist at the Lab, giving suggestions about practical optics at the weekly group meetings, and, again, this looked like an opportunity to bring up some subjects that interested me, as well as fleshing out the magazine. Because this was the Spectroscopy Lab, I always tried to angle my pieces toward that aspect of optics. The pieces I was most satisfied with were the ones about Balmer and the fitting of the hydrogen spectrum, and the one about Thomas Young and his near-miss on the topic of the supernumerary rainbow.

In these essays I have tried to follow in the path of those science essayists whose work I have enjoyed the most—Stephen Jay Gould, Willy Ley, Lyon Sprague de Camp, Isaac Asimov, and James Burke. Like most subjects, Optics is taught in a rather linear fashion, in which the subject is developed out of common observations and basic principles, proceeding from the more easily grasped topics and building up knowledge and concepts to provide a framework by which the student can understand more complex ideas, and is eventually enabled to read and understand contemporary articles. This is a reasonable way to proceed. It is often repeated for upper-level courses, giving the students more depth of understanding on various topics, but generally proceeding in the same way.

What it also does, as many have pointed out (including my roster of favorite essayists above) is to present a somewhat stilted view of history. Knowledge seems to invariably increase and grow in the correct direction, with periods of haziness but not definite error. There are no Dead Ends, and researchers seem to invariably Get Things Right. They seem to come up with intricate derivations, non-obvious experiments, and the most esoteric insights with minimal effort. Except for a few important or especially prolific scientists (Willebrord Snel, Isaac Newton, James Clerk Maxwell), most people who work in the field are invisible. Those we do know of, we seem to know only because they made one momentous discovery to which they attached their name, then disappeared into history. This is an almost inevitable result of the way the topic is taught. Most people do not study the history of science, or do not dip into the original research papers or the writings of those who work in the field, and thus do not learn about the rich and varied tapestry that is the True History of Science.

Thus, everyone knows about Willebrord Snel and his Law of Refraction. General knowledge may also extend to the fact that this law is called Descartes's Law in France. But in general the earlier history of discoverers and almost-discoverers of the law is not known because it doesn't tell the student anything more about the physics of the situation.

You could argue that this history of false starts, failures, and otherwise unattached names is of little use to the freshman student who just wants to propagate his light ray through an interface, but knowledge that such a backstory to the most basic of optical laws would teach him or her that the idea was by no means obvious, that the concepts themselves had to be identified and defined, and that the law really was discovered more than once. We do not have to rely on rare geniuses to advance our philosophies, but can all contribute to the endeavor. We should, in fact, if we wish to accomplish anything.

I have tried to make these essays accessible to anyone interested in the history and science of optics, without requiring a great deal of special knowledge. The original pieces had to be very brief to meet the requirements of publication in a magazine with limited space. I have taken advantage of this opportunity to present them in book form to extend them, adding background and explanatory material. The pieces that originally appeared in *The Spectrograph* have had additional material added so that it is accessible to an audience not made up of spectroscopists. I think that they should be comprehensible to a lay audience, although they may want to consult a book on optics or the Internet every now and then to check on the meaning of some concept in optics. I should also add that my wife (and editor and critic), who is not an expert in Optics or of Mathematics, informs me that if the reader isn't familiar with these subjects either, they can simply skip the indented portions containing any math and can simply resume their reading on the other side of them. It will all still make perfect sense. As someone who labored over the math, I think I should feel troubled by this, but if it means the work is more accessible, I'm in favor of the advice.

The book is broken down into three sections which the subjects addressed seemed to fall into. *History* concerns topics of historical interest—early uses of Optics, that article about Claudius Ptolemy and the Law of Refraction, the naming of the colors of the spectrum, a brief examination of the work of Robert Bunsen *besides* his eponymous burner, or of the underappreciated George Christoph Lichtenberg, or of Thomas Young's *other* work promoting the wave nature of light, besides *his* eponymous experiment.

Weird Science covers unusual events and effects and how they relate to other things. Why *is* the sun perceived as yellow, if sunlight is the very definition of White Light? How do retroreflectors work (there are more retroreflectors in heaven and earth than are dreamed of in your philosophy)? There are lasers you can eat, and lasers driven by fireworks, and lasers that seem to exist without being carefully aligned.

Pop Science is about popular perception of optics and topics where optics is relevant to some nontechnical topic. I have extensively revised and extended the section on the History of Ray Guns because it is one place where pop culture, in the form of juvenile fiction, comic books, radio, television, and the movies have produced their own history and image of a concept that is the province of optics without much input from optics, only to have a real device (the laser) fall into that pigeonhole, with attendant misconceptions. I look into these further in the Pop history of the Laser and in Popular Misconceptions in Optics. The Tractor Beam is another concept that has recently moved from the realm of wish-fulfilling fiction to the realm of possibility. Other pieces in this section look at how a patent could have been granted to the *idea* of using a laser pointer to play with your cat, a pop myth about camera-less imagery, glowing gems in story and in fact, and games that use light, among others.

I'd like to thank a great many people for their help in putting this together. First and foremost is my wife, Jill Silvester, always my first critic. I'd like to thank the people at *Optics and Photonics News* and the Optical Society of America, including my editors Lisa Rosenthal and Christina Folz, as well as Marko Batulan, Alessia Kirkland, Kathryn Amatrudo, and Hannah Bembia. I also owe Groot Gregory for suggesting my name as a Contributing Editor in the first place. My thanks to them all for their suggestions and revisions and for the illustrations they provided, and for allowing me to extend and publish these pieces in book form.

I also want to thank Professor Michael S. Feld, director of the George Harrison Spectroscopy Laboratory (and my freshman thesis supervisor at MIT), who encouraged and published the essays in *The Spectrograph*. To my great sorrow, he died in 2010, and so did not get to see this completed. My thanks also go to Ramachandra Dasari of the Spec Lab, and to Zina Queen, Vincent Russo, and Charles Holbrow.

PART I

HISTORY

{1}

Ancient Optics

MAGNIFICATION WITHOUT LENSES

Lenses do not appear to have been in widespread use as magnifiers or vision aids in the ancient world. They have rarely been found in the ruins of cities such as Pompeii or Herculaneum, which were buried suddenly and unexpectedly, with their household devices intact and in place. There are few mentions of them in the works of ancient philosophers and scientists, such as Euclid and Claudius Ptolemy. Seneca discussed atmospheric optics and the uses of curved mirrors, but his sole observation on refractive optics is that letters can be seen magnified when looking through a glass globe filled with water.[1] There have been glass globes found which can plausibly have served as magnifiers, when filled with water.[2] Aristophanes, in *The Clouds*,[3] mentions the use of a druggist's "stone"—probably a lens—to kindle fire. Most other authorities are silent. There don't seem to be any references to glass or crystal lenses for magnification or for correcting vision.

At the same time, optical researchers, historians, and archaeologists have noticed the existence of works of art with incredibly fine detail in the Classical World. Optical engineer (and optical historian) Keigo Iizuka mentions microscopic cuneiform in Mesopotamia.[4] George Sines and Yannis A. Sakellarakis[5] observed that there are many minute, detailed Roman gold-glass portraits. Pliny noted the existence of such tiny ancient works of art as an account of the *Iliad* written on a tiny scroll that could be rolled up and fit into a nutshell, or a tiny chariot small enough to be covered by a fly's wings.[6]

Did engravers and artisans use lenses to help them create these diminutive works of art? If so, one would think that there would be more mention of them. There have been discoveries of ancient lenses in one or two workshops, but no literary descriptions or mentions.[7] Lenses were certainly not common. An Assyrian text from 1800 B.C.E. indicates that the cost of a lens is about the same as that of an ox. Lens grinding and polishing are not trivial skills; they require polishing surfaces and grits of various size. In addition, there is the question of material—clear rock crystal was one of the most desirable materials, having a clear appearance. Much ancient glass is colored, has included material, and bubbles that are not easy to "fine" out. One thing in favor of those water-filled glass globe magnifiers is that a thin glass container filled with clear water was probably the best quality large-size magnifier material there was, at least until the water started to get algae and things growing in it, and had to be replaced. Perhaps the art of making and grinding lenses was maintained as a sort of guild secret, passed down through the societies of artisans and not shared with outsiders. That could explain the ancient silence on the subject.

I suggest that there is an alternative to these highly expensive, difficult-to-obtain, and (in the case of water spheres) breakable and perishable magnifiers. Something that was easily manufactured without special skill or materials and able to withstand everyday handling and moderate abuse without ill effect. I refer to the modest *pinhole*. A simple small hole placed in an appropriate material, if placed close to the eye, will restrict the angles of incoming rays and will provide a sharp image, at the cost of illumination. Many of us discovered this on our own, using randomly available small apertures, such as holes in buttons or punches in small buckle straps. Certainly the almost instinctual reaction of the myope is to squint, thereby using his own eyelid and brow as a makeshift aperture. A pinhole does the same more efficiently and with less strain.

One standard eye test, in fact, used by ophthalmologists is to place a pinhole in front of a patient's eye, then ask him or her to read a chart. This verifies that any eye problem that this corrects is an error of the eye's refractive system and not a defect in the retina or other portion of the eye. For a long time, until modern corneal sculpting methods were developed, pinhole "lenses" were the only means of correcting vision problems due to misshaped corneas, such as with *kerataconus*.

These pinholes work by "stopping down" the aperture. They minimize pupil aberrations and vastly increase the depth of focus. In order to work, the pinhole must be smaller than the pupil. It need not be extremely small.

It is a relatively easy matter to punch or drill such a pinhole in a variety of durable materials—leather, pottery, stone, seashell, wood, bone, horn, vegetable husks, vellum, metal, or a great many other materials. The hole needn't be round, but it is a simple matter to produce very round holes with punches or drills, which were used to put holes in beads, metalwork, and the like. The shape of the item bearing the pinhole might be a circular "monocle" (in which case the device could be "worn" by holding it in place between the upper portion of the cheek and the brow, like a monocle), or it could be in a hand-held "lorgnette." It could, in principle, be in a form like modern glasses, but I suspect that such developments were far in the future.

I note that these devices could be used in two important ways—they could act in the same way as corrective lenses, enabling the myopic to see (see chapter 32, "Pinhole Glasses," for more details), or they could be used as effective magnifiers.

Human eyes can only be used to within a certain distance, the *Accommodation length*, that is typically about 7 cm for young adults with "normal vision." Comfortable viewing distance for strain-free observation is normally taken to be about ten inches or 250 mm, and the magnifying power of lenses are calculated by assuming that the virtual image obtained by using a lens is set at that distance. To see things closer, one must use the ciliary muscles of the eye to tense the lens of the eye and shift the effective focal length of the eye closer. Prolonged operation at this close distance requires exerting these muscles and tires the eye. But the advantage of bringing the object closer to the eye is that a given distance on the surface will subtend a greater angle, and you can see it in more detail. The closer you bring it, the more detail you can see (at some cost in strain), but there comes a point beyond which you can't keep the object in focus—that closest accommodation distance.

To see an object subtending a greater angle, our solution is to use a magnifying glass, which gives us a virtual image larger than the object, and at a more comfortable viewing distance. Even farther from the eye at this distance, the image subtends a larger angle than it would if held close.

An alternative, using no lens, is to use a pinhole as a restricting aperture close to the eye. This lets you bring the object closer to the eye than the accommodation distance and still have the object in focus. You therefore effectively have lens-free magnification—the item subtends a larger angle than it does at the close accommodation point.

I have constructed several of these, using shim stock and a pushpin to make the hole, and they work quite well. If I make a hole about 1 mm in diameter it lets in sufficient light to work by daylight or a candle. I have made disc "monocles" and handle-held lorgnettes, and with both types I can bring items up very close to the eye and see small details. The monocle type has the advantage of letting me use both hands freely.

These "magnifiers" can also be used by observers, not just artisans. It explains how someone could examine and appreciate those tiny works of art—otherwise there would not be much point in making them.

Have such items been found? It's hard to say. If they were, how would people know what they were? A paddle or a disc with a hole in it might be anything. There have been a great many discs with holes found, but these are always identified as buttons or decorations. In China, there are round polished discs of jade or other substances, often with holes in the centers, called *Pi* discs (or *bi* discs). Many of these are quite large, with large apertures, but some are small enough, and with small enough holes, to make serviceable "monocles." Moreover, with their smooth polish, they would be perfect for placing near the eye. Did they serve as rough-and-tumble magnifiers? Who can tell?

The earliest reference I've found to the use of pinhole viewers in the West is a reference to masks with holes in them, made by Paulus Aeginēta, a seventh-century physician and medical writer in his *De Re Medica Libri Septum*.[8] (The use of tubes was attributed to Aristotle much earlier, but tubes are not as useful or efficient as pinholes. See chapter 32, "Pinhole Glasses")

After trying this out, I came across earlier suggestions about ancient magnification. Jay M. Enoch, who has written much about the possibilities of ancient lenses, observed that a droplet of water or oil suspended in a minute hole in a piece of metal or some other material could have been used as a magnifier. This is basically Seneca's water-filled glass globe, with surface tension acting as the glass. I have seen the idea suggested as an easy-to-make microscope in the 1960s. If a strip of metal supported on each end by blocks or folds in the metal itself is used, it can be "focused" by pressing down on the center of the strip to move the droplet towards the object under study.

L. Gorelick and A. J. Gwinnett[9] in 1981 suggested that much of the tiny detailed work could have been done without any sort of optical aid—even the pinhole magnifiers I suggest—if it was done by older workers with myopia. For such people, the minimum accommodation distance is moved closer to the eye as the person ages, and so they can naturally bring objects closer to the eye and see more detail than younger workers. Ophthalmologist Yale Solomon responded to this article by suggesting that older workers weren't needed—that using a pinhole in the same manner I have described would provide the needed magnification. In reply, Gorelick and Gwinnett said that they thought this would be impractical, especially since holding both the pinhole and the item being worked on would be difficult. But the pinhole could be mounted in a disc and held monocle-style, as I suggest. Or it could be hard-mounted near the object being worked on, just as workers in miniature often have their work mounted next to magnifying glasses. The objection is certainly not fatal to the concept.

Notes

1. Seneca *Natural Questions* I.6.5, in *Seneca: Naturales Quaestiones, Books I-III*, trans. Thomas H. Corcoran ([1971] 1999), 56–59.

2. Robert Temple, *The Crystal Sun: Rediscovering a Lost Technology of the Ancient World* ([1999] 2000), 57–59, 89–90. I hesitate to mention this book, which has a vast amount of errors and misinformation in it, but every now and then Temple makes an astute observation, and I think this is one of them. I caution care with this book, which is somewhat rambling and is filled with wildly unlikely and unsupported assertions (such as that lighthouses were used as telescopes, that the Greeks regularly manufactured telescopes, and that there were workshops making highly sophisticated deliberately aspheric lenses in ancient Mesopotamia), but he occasionally makes a few good suggestions and follows it up with investigation in the library and museums—a highly frustrating and lengthy book.

3. Line 760.

4. Keigo Iizuka, *Engineering Optics*, Springer Series in Optical Sciences, 35 ([1985] 2008).

5. George Sines and Yannis A. Sakellarakis, "Lenses in Antiquity," *Am. J. Archaeology* 91 (2): 191–196 (April 1987).

6. Temple, *Crystal Sun* 77–79, citing Pliny *Natural History* Book VII, chapter 21, in *Pliny*, trans. H. Rackham (1969), 2:561–563.

7. Jay M. Enoch, "The Enigma of Early Lens Use," *Technology and Culture* 39 (2): 273–291 (April 1998).

8. William Smith, *A Classical Dictionary of Biography, Mythology, and Geography* (1891), 531. Available at http://books.google.com/books?id=MaQUAAAAYAAJ&pg=PA531&dq=

Paulinus+of+Angina&hl=en&sa=X&ei=B6qYT9HAKIPXoQGTtJzoCQ&ved=0CDwQ6AEwAA#v=onepage&q=Paulinus%20of%20Angina&f=false, accessed May 1, 2013. Barnes improperly identifies him as "Paulinus of Angina" and puts him in the sixth century. I note, in addition, that there were reportedly metal pinhole spectacles in India, although I suspect that these were really from Astana in China from the eighth century. See "Pinhole Glasses," chapter 32 in this collection.

9. L. Gorelick and A. J. Gwinnett, "Close Work without Magnifying Lenses?" *Expedition* 23 (2): 27–34 (1981) and follow-up with question by Yale Solomon and reply by Gorelick and Gwinnett, *Expedition* 23 (4): 15–16 (Summer 1981). I observe that Gorelick and Gwinnett did a lot of work on ancient drills and their uses in cylinder seals and other ancient artwork, using scanning electron microscopy to observe fine detail. Such drills would be perfect for manufacturing the pinholes I suggest.

References

Barnes, R.J. "The History of Slit and Pinhole Spectacles," *Australian Journal of Optometry* 15 (6): 34–38 (Sept. 26, 1932). One of the few references on such pinhole devices, but it traces the idea back no earlier than Paulinus in the sixth century.

Marmor, M.F. and J. G. Ravin, *The Eye of the Artist* (1996), chapter 4, section 2.

Mathurand, S.S. and R. D. Bahuguna, "Reading with the Relaxed Eye," *Am. J. Phys.* 45 (11): 1097–1098 (Nov. 1977).

Sayanagi, K. "Pinhole Imagery," *J. Optical Society of America* 57 (9): 1091–1099 (1967).

{2}

The Solar Weapon of Archimedes

Archimedes was a third century B.C.E. mathematician and engineer from Syracuse in Sicily. Only a few of his works have come down to us, but those seem remarkably ahead of their time. In *The Sand Reckoner* he outlines a system for writing very large numbers, much like our system of exponentials, and then goes on to calculate how many grains of sand it would require to fill the universe. His work *On Floating Bodies* gave us Archimedes' Principle that the weight of a body placed in water is lessened by the weight of water displaced. His work on the circle resulted in a value of π between 3 10/71 and 3 1/7.

Archimedes was also supposed to have created exotic weapons of war for Hiero II, the ruler of Syracuse, to use against the Romans during the Second Punic War. Of all the weapons used—Catapults, Siege Towers, Giant Cranes—the one that seems to have captured the most modern interest was a giant mirror or collection of mirrors said to have been used to burn enemy ships. Lucian of Samosota and Galen both mention this, as does the sixth-century architect of the church of Hagia Sophia (usually called Saint Sophia), Anthemius of Tralles, who says Archimedes used many mirrors. Proclus, noted as a mathematician, was also said to have used burning mirrors against Vitellius in 514 C.E.

Can the story possibly be true? The consensus has ranged back and forth through the years. An excellent historical survey is to be found in the works of D. L. Simms of the British Scientific Civil Service, who devoted several years of study to the topic. Galileo Galilei evidently believed that it was possible at one time, and that it was not at another. In the seventeenth century, Athanasias Kircher, who seems to have studied it only theoretically (and produced engravings illustrating the possible configurations), decided it was a possibility. Rene Descartes, having analyzed the situation mathematically, deduced that it was not. George Louis leClerc, Comte Buffon, experimented with his own battery of mirrors in 1747 and achieved impressive results. Using 40 mirrors measuring 8′ x 6′ at a distance of 66 feet, all mounted on swivels and arranged to reflect the sun onto a single patch, he was able to ignite a piece of tarred beech. With 98 mirrors at a distance of 126 feet, he burned a wooden plank (wood not identified), covered with pitch and brimstone. With 128 mirrors at 150 feet, he set fire to tarred fir, and with only 45 mirrors at a distance of 20 feet, he could melt tin. Regardless of what you think about the historicity of Archimedes' mirror, that's still an impressive feat for a solar mirror. (Isaac Newton constructed a "burning mirror" from multiple flat mirrors, as well, which he presented to the Royal Society.)

In the early nineteenth century, there was again an interest in this, and several people experimented with the idea, most notably François Peyrard in 1807. As an odd footnote to this, a British nobleman, Thomas Cochrane, the tenth earl of Dundonald, came up with an invention of great destructive power, which the British Admiralty kept secret as "too horrible for humanity." It was called "Dundonald's Destroyer." A century later someone, noting the interest in burning mirrors, became convinced that the Destroyer was just such an instrument.[1] As I have noted in chapter 36, "Zap!," Washington Irving equipped his Lunar Invaders in *Conquest by the Moon* (1809) with solar concentrators—it seems likely that he was inspired by the period interest in the weapons. His story was written after Peyrard's reports.[2]

The 1914 Italian silent film *Cabiria*, directed by Giovanni Pastrone, depicted the Siege of Syracuse. Of the many inventions of Archimedes, the only one depicted was the Mirror, with which he is shown destroying the sails of the Roman fleet. The mirror is depicted as a single paraboloidal instrument made up of smaller mirrors.

In 1973 Albert C. Claus of Loyola University started a chain of comments by suggested that Archimedes could have performed his feat if he had, instead of a single large mirror, a team of soldiers armed with mirrors, each of which could be independently pointed.[3] He even proposed one method of pointing each mirror in the array.[4] O. N. Stavroudis wrote back to say that the idea of using burning mirrors on one's enemies predates Archimedes and can be found in Aristophanes' comedy *The Clouds* (produced in 423 B.C.E.). Furthermore, all the sources for Archimedes' feat date from centuries after his time. If there was no credible account of the event, why should we believe that it happened?[5]

As K. D. Mielenz pointed out, it appeared that the experiment had been done before by Comte Buffon.[6] Furthermore, in 1973 a Greek scientist, Ionannis Sakkas, got a team of 60 soldiers together, gave each a mirror, and had them direct the sun's rays at a boat 160 feet away. They succeeded in setting the boat afire. The incident was reported around the world, in the London *Times*, *Time* magazine, and *New Scientist*.[7]

In response, D. L. Simms wrote critiques of the issue, concluding that it was not reasonable to believe that it was possible. His long article in *Technology and Culture* in 1977 is probably the definitive historical account, examining the history of claims about Archimedes, Archimedes' knowledge about such burning mirrors, and the state of philosophy at the time.[8] Simms acknowledged the work of Sakkas and Buffon, yet concluded that these simulations were irrelevant, since they represented a "static situation," and that it is difficult to adjust the focus properly under "battlefield" conditions. He concluded that a much more likely weapon to have been used was "firepots" filled with tenacious and flammable material. The account given by Galen actually is more in line with this explanation than with a reflective light weapon.

The definitive work on the physics of the situation is probably "Reflections on the 'Burning Mirrors of Archimedes'" by A. A. Mills and R. Clift of the University, Leicester.[9] They examine the optics of the situation in detail, and their results will be cited and examined further on.

Archimedes has shown up in textbooks on Adaptive Optics as a sort of Patron Saint of Adaptive Optics, with his mirror being claimed as the earliest recorded case of a "Rubber Mirror."[10] In 2002 in Osnabrück, Germany, 500 volunteers used mirrors measuring 45 × 45 cm to ignite a black cloth sail at a distance of 50 meters.[11]

Solar furnaces have been built for decades, with large ones dating back to the 1940s. The PROMES-CNRS laboratory (from *Procedes, Materiaux, et Energie Solaire*, French for *Processes, Materials, and Solar Energy*) at Odeillo in the French Pyrenees has 63 solar-tracking flat mirrors that direct the light onto a parabolic array of 9,500 mirrors with a focal length of about 18 meters, which can produce temperatures of up to 3,800°C (over 6,800°F). Certainly the power of concentrated sunlight can produce impressive results. There is no contemporary evidence that Archimedes (or Proclus, or anyone else) ever actually used solar mirrors as a weapon of war, but *could* they have been used successfully?

The issue was taken up by the Discovery Channel series *Mythbusters* in the United States on three occasions (as of this writing). As is typical of the program, they came up with a showy visual, going so far as to build a section of an ancient boat as a target for their first outing in 2004.[12] For this trial, they used a single mirror made up of multiple flat mirror segments, as depicted in the movie *Cabiria*. They used 300 individual mirrors covering 400 square feet of area, placed onto a large plywood backing. A depth gauge was used to adjust the tilt on each mirror segment. The resulting giant mirror focused sunlight into a focal area about 2–3 feet in diameter on the section of ship, placed 60 feet away. Although the spot got 200 degrees Fahrenheit warmer, it didn't get hot enough to burn, and they concluded that, in the words of the show, "The Myth Was Busted!" In addition to their failure, they cited the unlikely conditions—you'd need a sunny day to use the Solar Death Ray, and an enemy willing to attack under those circumstances.[13]

Was it a fair test? The use of the crude focusing mechanism meant that the beam was not optimally focused. My own experience with using purely mechanical means to set optical alignments is that the results aren't very trustworthy, unless a great deal of care is taken with the alignment. It is almost always easier, quicker, and less expensive to align mirrors optically. The large size and evident lack of alignment of the mirrors in this case attested that this was not a good alignment, and the power was not concentrated in as small an area as one would want it to be.

Certainly the *Mythbusters* rating of "Busted" didn't sit well with some people. MIT Mechanical Engineering Professor David R. Wallace duplicated Archimedes' experiments for his 2.009 class Product Engineering Processes[14] in October of 2005.

Professor Wallace did some almost literal "back of the envelope" calculations to see if it was feasible to burn a wooden ship with mirrors. He calculated that 3 square meters of mirror focusing down to one square foot ought to provide enough flux to ignite wood.[15] Like *Mythbusters*, they built a target in the shape of a ship using 1′ thick red oak and tested the feasibility using 20 mirrors that were one-foot square. For the real experiment, they went up on the roof of a campus garage and set up 127 mirrors at a distance of 100 feet and a "staff" of 10 people to adjust the mirrors. The boat was coated with *MinWax* protective wax, since it was believed that Romans used wax to "paint" their boats. Each mirror was adjusted to send sunlight to the same point on the ship, then covered. When all mirrors had been adjusted, they were uncovered at once. They were able to char the wood, then generate open flame.

The MIT class was invited to perform the test on *Mythbusters* on October 22, 2005.[16] The experiment this time did not duplicate either the 2004 *Mythbusters* experiment or the 2.009 garage-top experiment.[17] They used as a target an actual boat, floating in water in San Francisco. Six students from the course went, along with Professor Wallace. They used 300 mirrors made of phosphor bronze, one-foot square, which were polished with *Brasso* metal polish They were set in a 110-foot long, four-tier high array whose center was 150 feet away from the target. Fourteen people worked to align the mirrors (Wallace remarks that "Adam [Savage] was an awesome mirror-aiming machine.") After uncovering all the mirrors at once (as at MIT), there was a bright spot on the side of the boat, which Wallace estimated as 4 to 5 feet in diameter, larger than they'd hoped (he says that about 30 percent of the mirrors were not flat, giving larger spots than anticipated). They repeated the aiming, getting a smaller spot (3–4 feet diameter), and were rewarded with glowing and smoldering embers along the side of the ship, but no flame. They tried to set fire to the sails, but no charring was observed. Wallace attributes this to the thin sails rapidly losing heat to the San Francisco winds.

They switched to silvered mirrors, getting similar results, then brought the boat in to 75 feet. A flame broke out near one of the scuppers, and they ultimately made a 10-inch hole in the boat. The experiment was abruptly terminated when a wind gust blew over several of the mirrors, shattering them. Wallace was generally pleased and thought the difficulties in this case probably were due to the moisture in the wood. In conversations with me, Professor Wallace said that the wood of the boat was not merely wet, but completely saturated with water, and that the ship had, until a few weeks before, been completely submerged.

Mythbusters judged this experiment "Busted" yet again, observing that it wasn't until they had cut the distance in half (and gone to modern mirrors) that ignition was achieved. They also listed the features which made the story unlikely.[18]

Mythbusters addressed the problem one more time on a special "Presidential Challenge" 2010 episode introduced by President Barack Obama. The episode, number 157, aired December 8. This time they got a crew of 500 volunteers (students), and armed each with a double-sided Mylar mirror, each of which appears to be 3′ × 6′, similar to the size of the shields used by Sakkas back in 1973. This time they used netting, placed in front of the volunteers, as an aid to aiming—they could use the spot on the netting as a guide to pointing their mirrors. At a distance of 30 meters, they failed to set the boat on fire, although they did raise the temperature to 230°F. Afterwards, they pulled the boat in closer, and "captain" Jamie Hyneman threw tennis balls at the mirror-bearing "soldiers." No flame was achieved, and they again declared the myth "Busted." Jamie Hyneman did say that the mirror weapon was not useless, however—the extremely bright and concentrated light did blind and disconcert anyone in the boat, and that was a nontrivial effect.[19]

So, how is it that Sakkas, Wallace, and the Osnabrück Germans claimed success, while *Mythbusters* (not to mention Descartes and Simms) declare the feat not possible? Leaving aside the question of whether Archimedes or Proclus actually did try this, shouldn't it still be possible? After all, one can produce awesome amounts of energy concentration with solar mirrors, as was demonstrated on the BBC2 television show *Richard Hammond's Engineering Connections* where he visits Keck Observatory. They covered a parabola with not-very-smooth aluminum foil and were able to set fire to a wooden boat.[20]

To answer these, we need to look a little closer at the physics and optics involved. The article by Mills and Clift does an excellent job of this. Many others writing about the subject have covered some of the less obvious details of the situation, but Mills and Clift have covered just about all of them in one place. Recognizing that there certainly is enough solar flux at the earth's surface to set fire to wood if enough of it can be concentrated, they go further to point out that the finite size of the sun in the sky produces a diverging beam. The sun occupies a bit over half a degree, so the rays diverge from a perfectly flat mirror at a half-angle of 16 minutes. If you use a round mirror of radius **r**, therefore, a distance **L** away from the mirror the illuminated spot will be larger, with a radius of

$$r + L^*\tan(16') = r + 0.00465^*L$$

Of course, if the mirrors are slightly convex, or irregular, the spot could be even bigger, and this accords well with the *Mythbusters* 2006 trial. (At 100 feet, as in Wallace's first attempt, the spot would have been about 60 cm across, only 2 feet.) The obvious fix, if you were in a situation like this, would be to make sure your mirrors are concave. In that case, they would focus the light and could compensate for the divergence the half-degree size of the sun introduces.[21]

The size of the spot produced by a concave mirror of focal length **f** is

$$2^*0.00465^*f = 0.0093^*f = f/107.4$$

This implies that there is an advantage in using the smallest possible focal length, in order to get the smallest spot, and in using as large an area as possible. For a short focal length and a large effective diameter to the "parabola" made of your multiple mirrors, you have a small f/number. As Mills and Clift point out, those powerful solar furnaces that produce impressively high temperatures, like the one in France, all have small f/numbers of 0.8 to 1.0. So, too, does the parabolic reflector that Hammond used on his BBC show. It's easy to concentrate a lot of light if you have your target very close to a short focal length mirror of broad extent. It's much harder in the more realistic circumstance where you only have a relatively short arc, pointing at a target very far away.

Of course, that is the circumstance in which it is most useful. One likes the idea of an insubstantial weapon that can project death at a distance, rather than actually throwing something physical, like a firepot. But the truth is that firepots, for all their low tech, are more effective. I'm irresistibly reminded of those who criticized laser weapons, especially during their early years. It often seemed that the difficulty of producing enough power at a distance required such a large device, with such large optics, that they said it would be more effective to simply drop the laser on the enemy than to fire it.

As Mills and Clift point out, a 1-foot diameter mirror with a 107.4-foot focal length would produce the same solar flux at 107.4 feet (assuming no reflection losses) as there is at the mirror itself. There is thus no concentration. But it is not true that there is no advantage—you have that flux now positioned on a target 107.4 feet away, and that flux is higher than you would get by simply using a flat mirror. Moreover, if you use multiple mirrors at 107.4 feet, you will multiply that flux. When they estimate the number of mirrors needed, Mills and Clift "stacked the deck" against Archimedes' mirror by assuming the use of plane mirrors, whereas by using concave mirrors one would significantly reduce the number needed.

The question of aiming is important, as well. The object for the soldiers is to direct the light from the sun onto the target of interest—and not merely onto the target, but, ideally, onto the same part of the target, to within a small margin of error. It's not immediately obvious how to do this—one does not casually bisect the angle between the sun and one's target and align the mirror perpendicular to this line. Adam's mechanical gauge method of directing the mirrors did not work well. When they tried getting multiple aimers to simultaneously adjust their reflections onto the same spot, it was difficult for people to tell which reflections were theirs. The MIT method of masking all mirrors but the one being pointed works somewhat, but it takes time to move each mirror, and the sun will move "off target" in the time needed to get multiple mirrors adjusted. Moreover, in a shootin' war one wants a real-time solution by which each soldier can adjust their own mirror. *Mythbusters* tried to do this in their third attempt with the "aiming net" stretched in front of the mirrors.

In fact, ways to point a mirror so that sunlight can be reflected onto a desired target have long been known. In 1942, during the Second World War, L. L. Young of the National Bureau of Standards and W. M. Potter of General Electric devised methods for aiming heliographs, intended to be used by pilots shot down and awaiting rescue, and used for signaling kits manufactured by GE. These methods are described in two articles by R. S. Hunter and published in the *Journal of the Optical Society of America* in 1945 and 1946 (after the war was over, but better late than never).[22] Of the several methods described, one resembled the "netting" approach used by *Mythbusters*. But a more effective approach was the one presented in those articles and later suggested by Albert C. Claus in his *Applied Optics* article of 1973. It is also described by Mills and Clift. Science fiction author Hal Clement (Harry Clement Stubbs) described it in his 1957 novel *Cycle of Fire*.

In the Young/Potter heliograph method, one uses a double-sided mirror with parallel faces and a hole drilled in it. You hold the mirror up in front of your face, relatively close to it. The sunlight falling through the hole illuminates the user's face. He looks at this spot in his reflection in the back surface of the mirror, and he looks at his target through the hole in the mirror. He then adjusts the mirror (tilting in two dimensions) until the reflected image of the spot on his face disappears into the hole. The mirror will now be oriented to send the sunlight toward the item centered in the hole.

Mills and Clift agree that it would work, but suggest that the smoldering patch (which is described as something similar to what Wallace saw) could be easily extinguished. Simms's criticism is that this doesn't guarantee that all the men will be aiming at the same thing. This seems to me to dismiss military training—if armies are good at anything, it's coordinating the actions of large groups of men. Provided the commander can direct the men to the same portion of the same target, they ought to have success. A more relevant criticism is whether the aiming can be directed well enough to overlap and maintain it. Simms doubted that ships bobbing on a water surface could be "followed" well enough by the aimers. But the experiences of Sakkas, the 2002 Osnabrück example, and the 2006 *Mythbusters* test—all done on ships floating on water—seems to show that this is not necessarily an insurmountable obstacle.

So the operation is certainly difficult, but there seems no need for the opponents to specify circumstances that would seem to make it more difficult than it really needs to be. Mills and Clift's dismissing concave mirrors, and Simms's dismissal of coincident aiming and invoking bobbing ships seem to be exaggerating the problems involved.

I thought to compare the results for the different configurations used, assuming Mills and Clift's figures (table 2.1). The solar flux at sea level is 0.93 kW/m^2. The flux needed to ignite wood, according to Simms, is 29 kW/m^2, but he suggests calculating using a figure twice as large to allow for misfocusing, errors, etc. (Wallace's calculations, by the way, assumed a value of 16 kW/m^2.) I assumed perfect reflection from each mirror, but used their areas, numbers, and distances from the targets. I then calculated the flux per unit area, assuming perfect alignment.

The results were interesting. Buffon's first experiment, using only 40 mirrors at a distance of 20 meters produced a flux of only 16.6 kW/m^2 (we will assume these units for all subsequent results), and he set tarred beech afire with this. One

TABLE 2.1 Comparison of Duplication of Archimedes' Mirror and Expected Optimal Flux Produced

Mirror Size (sq m)	Mirror Distance (m)	Number of Mirrors	Reference and Notes	Area Target Using Mills and Calculations	Flux at Target if all Lined Up
0.0929	30?	127	MIT one-foot sq mirrors burn wood in 10 minutes	0.197398	55.58524
1.6 estimate	48.8	60	Ioannis Sakkas 1973 set fire to ship	2.22556	40.11575
0.035	20	40	Buffon 1747 tarred beech set afire	0.078446	16.59732
0.035	38.4	98	Buffon 1747 tarred and brimstoned plank set afire	0.133695	23.85958
0.035	45.7	128	Buffon 1747 tarred fir set afire	0.15967	26.09374
0.035	6.1	45	Buffon 1747 melted tin	0.046418	31.55579
0.124	18.3	300	Mythbusters 2004 Single quasi-parabolic made of square mirrors	0.191171	180.9686
0.0929	45.7	300	Mythbusters 2006—MIT re-creation better aiming	0.267599	96.85788
1.6 estimate	30	500	Mythbusters 2010—aiming by individuals	1.97237	377.2111
1	50	440	Calculation only, minimum for burn	1.519056	269.3778
0.2	50	500	2002 Osnabrück, Germany set sail afire	0.462011	201.2941

must assume that the required flux is low when you tar the beech because this is much lower than Simms's values and is just about Wallace's estimate. The flux for his other cases was higher, but, even when he was melting tin (the highest flux he attained) it was 31.5, well below Simms's "practical" value.

Sakkas's 1973 successful ship-burning (for which I had to estimate the mirror sizes from the photographs, since he did not give them) came to 40, and Wallace's MIT rooftop results to 56, yet both of these produced open flame. The successful Osnabrück trial, using 500 mirrors at 50 meters, would have produced an astonishing 200 kW/m² if perfectly aimed—plenty, even for Simms's most pessimistic assumptions. The Mills and Clift result for the theoretical system they imagine is even higher at 270.

How about the *Mythbusters* results? If their 2004 single mirror had been well-aligned, it ought to have produced 180 kW/m² (using mirrors with an area of 0.124 m² on average at a distance of 18 m, the area on target would have been 0.19 m²). The *Mythbusters* 2006 flux would be a smaller 97 (although that still ought to work, being higher than the 56 Wallace obtained on the rooftop of the MIT garage), but the third *Mythbusters* attempt should have brought the highest flux of all—377 kW/m². (It was closer than the German attempt with its 500 mirrors, and their mirrors were much larger, so the overlap, even with mirrors moving around, ought to have been greater.)

Pretty clearly, all the mirrors were not lined up perfectly, and all things were not equal. Certainly the aiming in the first *Mythbusters* test was not as good as it should have been. Looking at the efforts of the aimers in their third try, I suspect that it was not as good either. It's not clear how much training they got and how well-coordinated their efforts were. The angular error introduced by the close-by netting would be greater than the error produced by the Claus-suggested method of Young and Potter. The reality of water-soaked wood and wind-moderated cooling would also play a part.

I can't definitively say that using concave mirrors (chosen to match the range, ideally, although even if this is not the case, this would reduce the spot area and increase the flux) with a large number of mirrors and well-trained and coordinated "soldiers" using the Young/Potter aiming scheme would work, but it hasn't been tried, and I'm reluctant to say that an Archimedes-type scheme wouldn't work. I observe, by the way, that really large mirrors aren't needed, and the only advantage they give is the larger probability of overlap—they do not increase the flux per unit area. For a test where you're coordinating the actions of a large number of people, and you want them to swing their mirrors to the same target in real time, there's an advantage in keeping the mirrors light and small enough to allow fine positioning and quick response. You ought to be able to get good results with mirrors a foot to a foot and a half in diameter.

Certainly it would take its toll on people more easily than on the boat itself. The flashing concentrated light would distract and dazzle anyone it was aimed at and perhaps temporarily blind them. Jamie Hyneman made this observation on the third *Mythbusters* attempt, and Mills and Clift had suggested it in their 1992 paper. Even before that, Arthur C. Clarke suggested this possibility in his short story "A Slight Case of Sunstroke," which appeared in the September 1958 issue of *Galaxy Science Fiction* (and was subsequently collected in his 1962 collection *Tales of Ten Worlds*). In this story, angry fans at a soccer game use the glossy covers of their game programs as mirrors to direct and concentrate sunlight onto an unpopular referee, with devastating results.[23]

This being the age of the Internet, there are plenty of individuals unaffiliated with institutes of higher learning or production companies for television who can run their own experiments and post the results online, often in highly dramatic *YouTube* footage. As I've said before, the cyberspace that these Internet sites exist in is somewhat less ephemeral than a telephone call, so they may be gone by the time you read this. Of course, the Internet being as mutable as it is, they may be replaced by other videos or descriptions[24, 25]

Notes

1. Edgar Stanton Maclay, "'Burning-Glasses' Dundonald's Destroyer?" *The North American Review* 201 (712): 434–438 (March 1915). The device showed up shortly thereafter as a solar weapon of destruction in Mary Raymond Shipman-Andrews's novel *Joy in the Morning*, and some have claimed that the novel is the source of *Dundonald's Destroyer*. But the novel came

out in 1919, four years after Maclay's article and five after George Harvey's "Dundonald's Destroyer" in the Nov. 1914 issue of *North American Review*. Actually, Dundonald's invention was very real and was a poison gas. Its identity was revealed in 1924. See Charles Stephenson, *The Admiral's Secret Weapon: Lord Dundonald and the Origins of Chemical Warfare* (2006).

2. As well as a pamphlet from the end of the eighteenth century and published in London, cited by Maclay. Irving was in Europe from 1804 to 1806.

3. Albert C. Claus,"On Archimedes' Burning Glass," *Applied Optics* 12 (10): A14 (October 1973).

4. In 1942 L. L. Young of the National Bureau of Standards and W. M. Potter of General Electric developed the method of aiming a heliograph described by Claus in his article. It was incorporated into signaling kits manufactured by GE that were intended for downed pilots. It's described in two articles by R. S. Hunter: "Letters," *J. Optical Society of America* 35: 805 (1945); "Heliographic Signaling Mirrors," *J. Optical Society of America* 36 (2): 110–115 (1946).

5. O. N. Stavroudis,"Comments on: On Archimedes' Burning Glass," *Applied Optics* 12 (10): A15 (October 1973).

6 K. D. Mielenz,"That Burning Glass," *Applied Optics* 13 (5): A14 (February 1974).

7. Excellent site with many references: http://www.mlahanas.de/Greeks/Mirrors.htm, accessed September 22, 2012.

8. D. L. Simms,"Archimedes and the Burning Mirrors of Syracuse," *Technology and Culture 18* (1): 1–24 (January 1977); see also D. L. Simms,"More on That Burning Glass of Archimedes," *Applied Optics* 13 (5): A16 (February 1974).

9. A. A. Mills and R. Clift,"Reflections on the 'Burning Mirrors of Archimedes,'" *Eur. J. Phys.* 13: 268–278 (1992).

10. Just a few of these references citing Archimedes as a father of Adaptive Optics are: J. W. Hardy, "Active Optics: A New Technology for the Control of Light," *Proc. IEEE* 66: 651 (1978); J. E. Pearson, in *Conference on Laser and Electro-Optical Systems*, San Diego, CA (1980); both cited in Robert K. Tyson, *Principles of Adaptive Optics,* 2nd ed. (1997), 3.

11. Cited in the same reference as note 7 above.

12. *Mythbusters* pictures, Season 2, Episode 16, first broadcast September 29, 2004: http://dsc.discovery.com/fansites/mythbusters/photogalleries/season_02/season_02.html, accessed September 22, 2012.

13. More *Mythbusters* pictures: http://dsc.discovery.com/fansites/mythbusters/photogalleries/season_02/season_02_02.html, accessed September 22, 2012.

14. I must explain that designation. Although it might seem to fit in with peoples' prejudices about MIT, it must be admitted that things there tend to be designated by numbers rather than names. Some of the buildings at MIT still lack actual names, and even those that have them are much better known by their numbers. Classes likewise are better known by their numbers. The part before the decimal denotes the department—departments are also known by their numbers—and the part gives the course number. At MIT the beginning calculus class isn't *Math 100*. It's *18.01*, since Course 18 is the Mathematics department. Course 2, similarly, is Mechanical Engineering—MIT has nine distinct engineering departments—and so 2.009 is a course in the Mechanical Engineering department, and an undergraduate one, judging by the low number.

15. His web page, which includes a scan of his "envelope," is online at: http://web.mit.edu/2.009/www/experiments/deathray/10_ArchimedesResult.html, accessed September 22, 2012.

16. The episode aired January 21, 2006.

17. Wallace's account is online at: http://web.mit.edu/2.009/www/experiments/deathray/10_Mythbusters.html, accessed September 22, 2012.

18. One account is given at: http://kwc.org/mythbusters/2006/01/episode_46_archimedes_death_ra.html, accessed September 22, 2012.

19. One site for this is: http://dsc.discovery.com/tv-shows/mythbusters/videos/presidents-challenge.htm, accessed October 14, 2012.

20. http://natgeotv.com/uk/engineering-connections/galleries/keck-observatory/1 http://www.bbc.co.uk/programmes/b00j2jlg, accessed September 22, 2012. Show aired March 14, 2010.

21. As a result of this divergence due to the angular size of the sun in the sky, the shape of a spot of light reflected from a mirror will become circular as one moves farther away from the mirror. If the soldiers employed by Archimedes used small mirror tiles, such as those used in mosaics, the result at a distance of 50 meters would be a circle about 45 cm in diameter.

22. See note 4. The same method of signaling was used in Hal Clement's science fiction novel *Cycle of Fire* (1957).

23. It subsequently appeared in his collection of short stories *Tales of Ten Worlds* (1962). A web page performing the calculations is here: http://zebu.uoregon.edu/~dmason/probs/therm/dyn/sunstrok/sunstrok.html, accessed September 22, 2012.

24. As of October 2012, these videos can be found at:

(a.) http://vimeo.com/19836043, accessed September 22, 2012.

Paul Bader demonstrates it with small mirrors in the actual harbor at Syracuse on a piece of cardboard. His spots appear to be lines, rather than square or spots, suggesting that his mirrors are actually somewhat concave cylinders.

(b.) http://www.youtube.com/watch?v=a9qk1loLjEs, accessed September 22, 2012.

GreenPowerScience spot, using multiple mirrors 95 feet away, pointing at a wooden cutout of a boat. They use concave mirrors to produce a smaller spot. Unlike the 2006 *Mythbusters* trial, they were able to set fire to their sail as well. They appear to have something like 30 mirrors, some 30′ x 40′, some appear to be 12′ x 12′, and others seem much larger.

(c.) http://www.youtube.com/watch?v=a8R5chBuHZg&feature=relmfu, accessed September 22, 2012.

In this video GreenPowerScience uses seven one-foot mirrors on a Honda, raising the local temperature to 300°F.

(d.) http://www.youtube.com/watch?v=-tkYCy-l990, accessed September 22, 2012.

Also from GreenPowerScience, using two 30′ x 40′ mirrors at 118 feet to set fire to black corrugated cardboard.

(e.) http://www.youtube.com/watch?v=F_yCPX8r_9s&feature=relmfu, accessed September 22, 2012.

Similar test using three mirrors at 20 feet on black and unblackened cardboard.

(f.) http://www.youtube.com/watch?v=MDkOaPp_6ug, accessed September 22, 2012.1990 video of students using coordinated mirrors to ignite (and explode) a box.

25. Archimedes' Mirror is described in Jearl D. Walker's *The Flying Circus of Physics* (1975), entry 3.76, on p. 65. Curiously enough, in this heavily footnoted book, no references are given. But in the revised edition "With Answers," published in 1977, some explanations are given on p. 247, but without citations.

{3}

Claudius Ptolemy's Law of Refraction

Emus tenui aqua tegitur et fracti speciem reddit...

"An oar is covered with shallow water and gives the appearance of being broken." That's Lucius Annaeus Seneca, noted Roman philosopher and author, writing in his book *Natural Questions* circa 63 C.E.[1] It's an early observation that light passing into water gets refracted and makes anything seen through the surface appear to bend. But it's not the earliest observation of that principle. Titus Lucretius Caro had expressed much the same thought in his philosophical poem *De Rerum Natura* (the title is sometimes grandiosely translated as *On the Nature of the Universe*, but it could as easily be rendered *The Way Thing Are*) about a century earlier.[2] Lucretius's poem was a rendering down of Epicurean philosophy, so the observation could easily be centuries older. It probably was—it's hard to imagine those seagoing Greeks not noticing the way the line of a submerged oar seems to "break" at the interface between air and water.

In the third century B.C.E., Ktesibios of Alexandria, possibly the first head of the Museum and Library there, performed an experiment demonstrating the effect in a slightly different way. A coin placed at the bottom of a cup would become invisible to a person who lowered their head until the lip of the cup just blocked their view of it. But when water was poured into the cup to the brim, the coin became visible again, as the light was bent at the interface between the air and the water.

This is the science of Optics as we usually expect to see it in the ancient world—a single observation, rather vague, quantitative rather than qualitative. That's not always the case. Some observations were surprisingly accurate and modern, seemingly far ahead of their time. They weren't, of course—they were part and parcel of a tradition of careful and exact measurements that is largely untranslated into modern English, not widely publicized today, and was largely forgotten before the Fall of Rome. They're well known to some Classical scholars and to historians of science, but not to the general public or even to most scientists. When you stumble upon one of these gems, like *The Sand-Reckoner* by Archimedes—which has a discussion of extremely large numbers and estimates the number of grains of sand needed to fill the Universe—you are astonished at its clarity and modernity. You

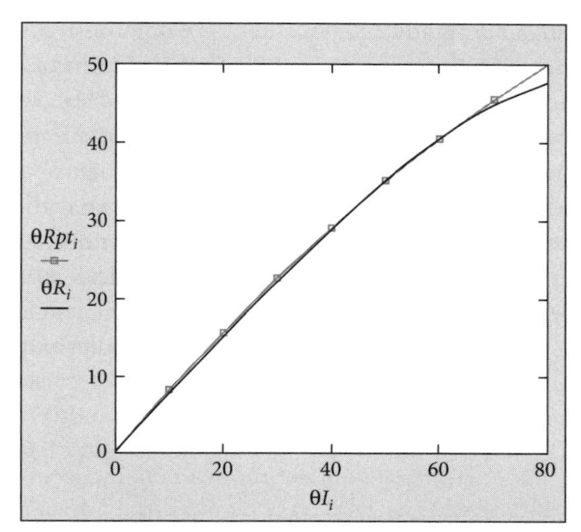

FIGURE 3.1 *Plot of Angle of Refraction vs Angle of Incidence at the interface between air and water as measured by Ptolemy (squares) and predicted by Snell's Law.*

wonder how this bit of knowledge could have been allowed to fall into obscurity in the ancient world.

One such piece of ancient science that I stumbled across is Claudius Ptolemy's determination of the Law of Refraction, which lifts these vague and dimensionless observations about refraction to the realm of quantified science. I had learned, as I think practically every optics student has, that the law of refraction was discovered by the Dutch mathematician Willebrord van Roijen Snell in 1621 and was first published by Christiaan Huygens in his 1703 book *Dioptrica*. Snell did not publish his law during his lifetime and was independently discovered by Rene Descartes. It is thus known as "Descartes's Law" in France. With this powerful law, Descartes was able to calculate the location of the rainbow. Newton was able to derive the laws of imaging with lenses. Because of the careful experimental work of Snell, Descartes, and others, modern optics took root in the seventeenth century.[3]

In recent years the history has become somewhat more complex. To begin with, Snell seems to have spelled his Dutch name with a single "L"—*Snel*. But he used his Latin cognomen of *Snellius* in his scientific correspondence, and it was wrongly assumed that he spelled his birth name *Snell*. Then there is the fact that Thomas Harriot discovered the relationship almost twenty years earlier, in 1602, but did not publish the result.[4] All of these efforts were long predated by Abu Sa'd al-'Ala' Ibn Sahl, whose book *On Burning Mirrors and Lenses* was published in Baghdad in 984. His drawings indicate the role of the sine of the angle in the law.[5–8]

But the mathematician, astronomer, and physicist Claudius Ptolemy had already performed a series of careful experiments in the first century C.E., long predating even ibn Sahl, to determine the rules of refraction. What's more, it was a quantitative study, and a startlingly modern one, performed just the way it is performed in many undergraduate optics labs. Ptolemy constructed a straightforward goniometer, marking off the degrees along the edge and placing the center at the interface between two different transparent media. He examined refraction at the interface between air and water, between glass and air, and between glass and water. He varied the angle of incidence between 10° and 80° in units of 10° (measured from the normal to the interface) and measured the corresponding angle of refraction. George Sarton[9] called this work "the most remarkable experimental research of antiquity." Ptolemy's work certainly does follow the model of experimental science we've been brought up to revere. He made his observation (that apparently "broken" oar, or something like it), hypothesized a cause (change of angle at the water/air interface), arranged an experiment in exemplary fashion, made his observations, and came up with a result. So why didn't history follow the rest of the script, giving us modern optics a millennium and a half earlier than we got it?

To listen to some of the critics, you'd think that Ptolemy *didn't* get it right at all. A look at Internet sites dealing with it imply that Ptolemy's numbers were off by quite a bit, and that his big assumption was that for small angles, the angle of incidence is proportional to the angle of refraction.[10] Wikipedia[11] accuses him of "confirmation bias." C. B. Boyer,[12] after presenting the hard evidence, notes that: "A closer glance at it, however, suggests that there was less experimentation involved in it than originally was thought.... As in other portions of Greek science, confidence in mathematics was here greater than that in the evidence of the senses, although the value corresponding to 60° agrees remarkably well with experience."[13]

Someone reading these critics would thus respond: "OK, Ptolemy had a good idea, but he didn't follow through well. He wasn't a good enough experimentalist, or he let his expectations rule the observations. He got the figure right at about sixty degrees, but not on the other values."

Only that's *not* correct. I submit, as evidence, Figure 3.1, which plots Ptolemy's results for the air/water interface alongside modern results. Far from that 60° data point being the sole intersection with reality, Ptolemy's data agrees with modern results to within about 0.5° for all values except 80°. (In fact, the best agreement isn't at 60°, but at 70°.)

We get similar results if we plot Ptolemy's results vs. modern results for the air/glass or the water/glass interface. (Ptolemy doesn't tell us what sort of glass, so I assumed a refractive index value of **n = 1.5**. Assuming that glass has an index of about one and a half is the first refuge of the Optical Engineer, who starts off playing with BK7 glass with its index of 1.55.) The modern law of refraction has the familiar form

$$n_1 \sin\theta_1 = n_2 \sin\theta_2$$

Here $\mathbf{n_1}$ and $\mathbf{n_2}$ are the refractive indices of the materials the light is coming from and incident on, and $\boldsymbol{\theta_1}$ and $\boldsymbol{\theta_2}$ are the angles of incidence and refraction, both measured from the normal to the surface.

All things considered, Ptolemy's results are pretty darned good. So why does everyone say they're bad?

One reason has to be that his work isn't widely distributed, even today, so most people haven't seen it firsthand. They're all repeating what they've heard from others. The figures for the air/water interface are reprinted in Boyer's book *The Rainbow: From Myth to Mathematics*, and you can find them on some Internet sites,[13] but his figures for the air/glass interface and the water/glass interface are harder to come by. I obtained them from A. Mark Smith's *Ptolemy and the Foundations of Ancient Mathematical Optics*.[14] Smith took them from his own translation of Ptolemy's *Optics*. This, as far as I can tell, is the first translation of this important work into English, and it appeared as recently as 1996.[15]

Several writers take Ptolemy to task for not having determined the correct relationship between the sines of the angles of incidence and refraction. This is somewhat ironic, since Ptolemy compiled what is essentially the first table of sines. But it's by no means obvious that it would be the sines of the angles that would be the function to have a linear relationship. Moreover, Ptolemy's tables did not give the sines of the angles directly, but rather the lengths of chords of those angles, measured on a circle having a radius of 60 units. The chord of an angle is proportional to the sine of half the angle, and this makes the relationship even less obvious.[16]

Furthermore, Ptolemy was very clearly taken by another feature of his tables of angles. In all three cases (air/water interface, air/glass interface, and water/glass interface) the second differences between the angles of refraction are constant. In other words, while the angle of incidence increases by 10° each time (10°, 20°, 30°, etc. all differ by 10°), the intervals between the angles of refraction are decreasing (8°, 15.5°, 22.5°, 29° differ by, successively, 7.5°, 7°, and 6.5°) The differences between the intervals is always −0.5°. Furthermore, Ptolemy found the second differences to be the same in all three cases—it's always −0.5°. This amounts to saying that the relationship between the angles of incidence and refraction, measured in degrees, is quadratic. Happily, the equations assume a relatively simple form. For air/water, the relationship is:

$$\theta_{refract} = \frac{33}{40}\theta_{incident} - \frac{1}{400}\theta^2_{incident}$$

For air/glass the linear coefficient becomes 29/40, and for water/glass it becomes 39/40. The quadratic coefficient is always (−1/400).

Two errors are involved—Ptolemy rounded his numbers, almost always to the nearest half of a degree. But not always, especially for the refraction corresponding to 80° incidence. He evidently saw the beginnings of a quadratic pattern and allowed it to shape his reported observations. As has been noted by others,[17] Ptolemy's reported angles of refraction aren't really "raw" data—they've been smoothed and massaged.

Smith suggests that Ptolemy was "conditioned" to see a pattern of constant second differences from his work in astronomy,[18] and so it's not surprising that he noticed what he thought was the same pattern in optics. If you use the correct values and Snell's law, of course, you find that the second difference is not a constant −0.5—the second derivative of the function varies considerably from 0 deg^{-1} at 0° to about −1.5 deg^{-1} at 70° for air/water.

We don't have Ptolemy's experimental values, of course—his figures represent his "best fit," but we do know what the refractive index of water is, and we have Snell's law. If we calculate what the actual values would be for the angle of refraction at 10°, 20°, 30°, and so forth, then perform a quadratic least square fit (with no offset term—we want the plot to go through the origin), then we obtain values very nearly equal to Ptolemy's—instead of a slope of 33/40 = 0.825 we get 0.836, and instead of a second order coefficient of −1/400 = −0.0025 we get −0.00288. Plotted out, the difference is very small and may simply be the difference between our decimal notation and what can be expressed as simply fractions with fairly round numbers.

Why did Ptolemy fail to discover Snell's Law? Smith says that Ptolemy wasn't trying to answer the same questions—he was not seeking to explain the radiation of light, but to explain the nature of sight.[19] But to explain the one is to explain the other. Ptolemy had great successes elsewhere in his work, why should he fail here?

It seems clear to me that Ptolemy felt he had reached the correct result and, having found the answer, pushed no farther. It was clearly not of central importance to him (despite the investment in time and research equipment), and, having satisfied himself with an answer, he moved on to something else.

Ptolemy's result may not have been perfect, but it was, as we show above, not a bad result. It certainly provided concrete numbers for anyone wishing to pursue a mathematical study of refractive optics. If we took a high school student or college undergraduate who had not heard of Snell's law and tasked him or her with making the measurements, then of coming up with a general law that fitted the observed results, it wouldn't be at all surprising if they came up with precisely the results that Ptolemy had. Clearly a plot of the refracted angle vs. the incident angle starts out as a linear function, but starts to depart from that. With a standard background in physics, one would naturally assume that a higher order term would be coming into play, and the obvious term to add is the next highest power, a quadratic, varying with the square of the angle.

Despite claims by those not fully familiar with Ptolemy's work, these results are not really all that bad. With the advantage of two millennia of mathematics and physics, we can see that the function

$$\theta_{refract} = \sin^{-1}\left\{\frac{n_{incident}}{n_{refract}}\sin(\theta_{incident})\right\}$$

ought to be expanded as a linear term followed by a cubic term in the next order, rather than a quadratic term.

$$\theta_{refract} \cong \frac{n_{incident}}{n_{refract}}\theta_{incident} + \frac{1}{6}\left(\frac{n_{incident}}{n_{refract}}\right)\left(\left(\frac{n_{incident}}{n_{refract}}\right)^2 - 1\right)(\theta_{incident})^3$$

If you only use the first two terms of the expansion, your error is less than Ptolemy's for angles of incidence between about 25° and 40°, but Ptolemy's formula actually has a smaller error for angles above 40°. So maybe it's not so bad to stick to his formulation. What would have happened if someone had actually used Ptolemy's law to formulate Geometrical Optics?

Well, the first thing you find is that the refractive indices don't come out quite right. Water's is 40/33, giving water a refractive index of 1.212 instead of 1.33. Glass has an index of 40/29 = 1.37 instead of 1.5. These results carry through to cause problems in later calculations. You can derive the system of paraxial optics with his formula and end up with the same results we have today, including the Lensmaker's Formula for the power of a lens (since the assumptions of a linear relationship between angles of incidence and angles of refraction was expressed by Ptolemy for small angles). But if you were to "plug in" the above values you'd find that the incorrect refractive index leads to an error in the final result. The predicted Power of the lens would be too low, the focal length too long.

In fact, one need not go to all that trouble. We can consider the problem addressed by Ktesibios—the fact that a ring placed at the bottom of a basin appears to be closer when the basin is filled with water.[20] Using very direct and simple paraxial optics, you calculate that the apparent distance of the ring from the water's surface is **(1/n)** times the distance it truly is. This would be a simple experiment to carry out, and the difference between an index of 1.21 and 1.33 would have been speedily noted.

The problem is, of course, that no one did the experiment. Ptolemy never followed through on his work, and no one else carried it on. If anyone had pursued this tantalizing lead, they would undoubtedly have discovered the inconsistencies. That, of course, is the sort of thing that is necessary for good science—peer review and testing of hypotheses. Ptolemy's experiment resembles less a modern research paper than it does an undergraduate experiment—performed once, printed, and not reviewed again. Had Ptolemy been forced to explain the discrepancies, he might even have produced the Law of Refraction as we know it, despite those cumbersome sines expressed as chords.

But Optics was not treated as well as other branches of mathematics and science until fairly recently. The fact is proven by the fragmentary state of our knowledge of ancient optics. We know that several books on the topic were penned, but we have few of them, and none of them complete. This indicates that few copies were made and transmitted. For Ptolemy's own *Optics* we have only a twelfth-century Latin translation of an Arabic copy, and we lack Book I and the end of Book V.[21] And, as I noted, an English translation has only become available in 1996. Ptolemy's ideas did not get fully corrected for the same reason they did not have a big effect in the history of optics—they were not widely enough spread, not well enough known.

As science historian Otto Neugebauer noted, though, Ptolemy's work had its greatest influence through the Arabic students of Optics, who knew of it indirectly through the work of Ibn al-Haitham (better known as Alhazen). But even here, it did not have full effect. Among the lost parts of Ptolemy's *Optics* is his explanation of the Rainbow. The Arabic Opticists Qutb al-Din al-Shirazi (1236–1311) and his student Kamal al-Din al-Farisi (circa 1320) both studied the rainbow, producing surprisingly modern results. They modeled the raindrop as a sphere of water and experimented with a glass sphere filled with water, following the beam of light as it refracted, reflected once inside, then refracted upon leaving the drop. They correctly explained both the primary and secondary rainbow in this way, and made the first observation of a tertiary rainbow. Only a few years later, a French-German monk named Theodoric of Freibourg was doing the same thing in Europe. His drawings have come down to us (al-Shirazi's and al-Farisi's, sadly, have not), and his pictures of a primary or a secondary rainbow could have been taken from a modern optics text. All three of these—al Farisi, al Shirazi, and Theodoric, observed that the ray of light bends as it enters the drop, and as it leaves. They knew, at least qualitatively, about the law of refraction bending light toward the normal as the light went from air into water, and away from the normal as it went from water into air.

Rene Descartes used Snell's law to imagine the trajectories of rays entering the drop at different distances from the center, and, noting that the rainbows occurred at a maxima or a minima in the angle of refraction, calculated for the first time the angle of the rainbow. If armed with Ptolemy's formula, al Farisi, al Shirazi, or Theodoric could have done the same 300 years earlier. Moreover, they would have gotten the correct result. Unlike the paraxial optics examples I noted above, which depend upon the values at small angles (where Ptolemy's insistence upon constant second differences led him into error), the rainbow angle is calculated upon angles where Ptolemy's results are almost perfectly correct (despite his fudging the data). The quadratic formula you get from Ptolemy's data predicts the rainbow to lie at 42° from the anti-solar point. The modern geometrical optics result, using Snell's Law, predicts 42.5°. [22, 23]

In other words, we came extremely close to having a correct theoretical explanation for the rainbow several centuries in advance of its actual discovery, and only missed it (arguably) because Ptolemy's book didn't quite last long enough or wasn't popular enough to reach the proper hands.

Notes

1. The quote comes from Book I, section 3.9. I quote both the Latin and the translation from the Loeb Classical Library edition of *Seneca: Natural Questions (Naturales Quaestiones), Books I–III*, translated by Thomas H. Corcoran (1971), 38–39.

2. Lucretius's observation of the apparent bending of the oar (and of the rudder as well) appears in Book IV, line 437. In the Penguin translation by R. E. Latham this is rendered as: "So much of the oars as projects above the waterline is straight, and so is the upper part of the rudder. But all the submerged parts appear refracted and wrenched round in an upward direction and almost as though bent right back so as to float on the surface" (1951, 144).

3. This sentiment is expressed, for instance, in Eugene Hecht's *Optics*, 2nd ed. (1987), 3: "... this was one of the great moments in optics. By learning precisely how rays of light are redirected on traversing a boundary between two media, Snell in one swoop swung open the door to modern applied optics."

4. A. Kwan, J. Dudley, and E. Lantz, "Who Really Discovered Snell's Law?" *Physics World* 15 (4): 64 (2003).

5. Roshdi Rashed, "A Pioneer in Anaclastics: ibn Sahl on Burning Mirrors and Lenses," *Isis* 81 (3): 464–491 (1990).

6. Roshdi Rashed et al, "Geometrie et Dioptrique au Xe siecle: Ibn Sahl, a-Quhi et Ibn al-Haytham," *in Collection Sciences et Philosophie Arabes, Textes, et Etudes* (1993).

7. Roshdi Rashed, "A Polymath in the 10th Century," *Science* 297 (5582): 773 (2 August 2002).

8. Sameen Ahmed Khan, "Arab Origins of the Discovery of the Refraction of Light," *Optics and Photonics News* 18 (10): 22–23 (2007).

9. George Sarton, *Introduction to the History of Science* (1927–47), cited in C. B. Boyer, *The Rainbow: From Myth to Mathematics* (1959), 61–62.

10. For instance: "Ptolemy formulated a very inaccurate version of the law of refraction, which only works when the light rays are almost normally incident on the interface in question. Despite its obvious inaccuracy, Ptolemy's theory of refraction persisted for nearly 1500 years." From Richard Fitzpatrick's "History of Geometrical Optics" (July 14, 2007): http://farside.ph.utexas.edu/teaching/302l/lectures/node125.html, accessed March 28, 2012.

11. http://en.wikipedia.org/wiki/Snell%27s_Law, accessed March 28, 2012.

12. Boyer, *The Rainbow*.

13. For instance: http://www.solutioninn.com/physics/electrodynamics/electromagnetic-waves/in-about-a.d.-150-claudius-ptolemy-gave-the-following-measured, accessed March 28, 2012.

14. Smith's book, published by the American Philosophical Society in 1999, is actually *Transactions of the American Philosophical Society*, Vol. 89, Part 3.

15. A. Mark Smith, "Ptolemy's Theory of Visual Perception: An English Translation of the Optics with Introduction and Commentary," *Transactions of the American Philosophical Society*, Vol. 86, Part 2 (1996).

16. See, for example, Otto Neugebauer, *A History of Ancient Mathematical Astronomy* (1975), Vol. 1, pp. 21ff.

17. Boyer says this, and so does Smith in his 1999 book. See also Otto Neugebauer's *A History of Ancient Mathematical Astronomy* (1975), Vol. 2, pp. 894–896.

18. See, for instance, James Evans, *The History and Practice of Ancient Astronomy* (1998), 333, for an example from Babylon.

19. Smith, *Ptolemy and the Foundations of Ancient Mathematical Optics,* 8. See also his exhaustive (and exhausting) article, "Ptolemy's Search for a Law of Refraction: A Case-Study in the Classical Methodology of 'Saving the Appearances' and its Limitations," *Archive for History of Exact Sciences* 26: 221–240 (1982).

20. Lucius Annaeus Seneca observed and noted this phenomenon, and tried to use it to argue his case for the rainbow. See his *Natural Questions (Naturales Quaestiones)*, Section I, 6.5, in Thomas Corcoran's translation in the Loeb Classical Library (1971), 54–57.

21. See Neugebauer, *A History of Ancient Mathematical Astronomy*, Vol. 2, pp. 892–893.

22. See Boyer, *The Rainbow*, 110–125.

23. See, for example, R. A. R. Tricker's *Introduction to Meteorological Optics* (year), 50–53. I leave the derivation of the formulas for higher order rainbows as an exercise for the reader. I note, with regret, that Ptolemy's results don't work so well for secondary rainbows, predicting that they lie at 35°, rather than the true value of 50°.

{4}

Antonio de Ulloa's Mystery

According to *Appleton's Cyclopedia of American Biography*, an encyclopedia of world events and leaders published in 1879, in 1779 Antonio de Ulloa y de la Torre Giral, Rear Admiral of Spain, Lieutenant-General of the Spanish Naval Forces, one-time governor of present-day Peru and future governor of the Spanish territory of Louisiana, was sent in command of a fleet to the Azores.[1] Along with him traveled sealed orders, to be opened upon his arrival there. These orders directed him to proceed to Havana, Cuba, and there take charge of an expedition to reconquer Florida for Spain.[2] This would have profound implications for the history of North America, and for the American colonies, now struggling to gain their freedom from Great Britain.

It would have, at any rate, were the officer in question anyone else. But Antonio de Ulloa was remarkable in being not only a naval officer and high royal official but also one of the most noted scientists of his day, a participant in the Spanish expedition to measure the arc of the meridian in 1736, describer of the geology of South America (including the first description of platinum), student of botany and observer of optical phenomena. He was a rare foreign member of the Royal Society of Britain. *Appleton's Cyclopedia* (1887–1889) says that

> entirely occupied with scientific observations, Ulloa forgot to open his sealed orders, and, returning to Cadiz after a cruise of two months, was arrested and tried by a court-martial in December, 1780, which acquitted him, but recommended him for land duty.

It's a great story, one of geopolitics combined with the ultimate absent-minded professor story. Unfortunately, like almost all such great stories, it apparently isn't true. No biography of de Ulloa published since 1900 credits him with such dereliction. The entry in the *National American Biography*, for instance, attributes the "charges of neglect of duty growing out of the loss of a Spanish ship and his failure to capture eight British merchantmen."[3]

The reason that the multitalented and posthumously maligned Antonio de Ulloa has landed in this book is that he was sent on an expedition to measure the arc of longitude in South America in 1735, and while in present-day

Ecuador he made a series of interesting optical observations. He is credited with the first report of the white rainbow. He also observed the Glory in the cloudbank surrounding his shadow in the valley below Mount Pambamarca. Both of these observations were reported in *Relacion Historica del Viage a la America Meridional*, written with his collaborator Jorge Juan and published in Madrid in 1748.

It was during the expedition to measure the arc in 1735 that Antonio de Ulloa, along with Don George Juan, visited the Paramos of present-day Ecuador and made a number of interesting optical observations. From the top of the Mount Pambamarca, de Ulloa observed the Glory, that series of tight rainbow-like circles seen around one's shadow on a cloudbank below. Today, the phenomenon is easily and frequently seen around the shadow of an airplane when one is flying, but in those pre-flight days one of the few places it could be seen was when one looked down into a cloudbank from the top of a mountain.

He also observed the colorless rainbow that occurs when the droplets are smaller than about 25 microns. Today such a rainbow is usually called a *fogbow*, an indication of the usual circumstances for observation, but an older name honoring its discoverer is the *Circle of Ulloa*. It was through this designation that I first learned of de Ulloa and his observations.

At the same time that he made these observations, de Ulloa also made a third, more mysterious one. He described it as a series of three arcs, intersecting at one point, observed by moonlight, and without color (see figure 4.1). Lynch and Futterman[4] translate the relevant passage:

FIGURE 4.1 *Antonio de Ulloa's illustration showing a Volcane, a Glory, and a Third Phenomenon.*

> On several occasions we noticed that in these Mountains the Arches were formed by the light of the Moon. I saw a quite singular one the fourth of April 1738, in the Plain of Turubamba at 8 o'clock in the evening; but the most extraordinary of all was observed by Don George Juan on the Mountain of Quinoa-Loma on the 22nd of May, 1739 at eight o'clock in the evening. These Arches were composed of no other color than white, and formed themselves in the slope at the top of a Mountain. The one which we saw was composed of three Arches meeting at a single point. The diameter of that in the middle was 60 degrees, and the thickness of the color white occupied a space of five degrees. The two other Arches were similar to that one.

De Ulloa and Juan oversaw the preparation of the engraving illustrating their work—they refer to it in the text—so the peculiar white arcs are clearly meant to convey what they saw in South America. The Fogbow and the Glory are clear enough, and represent well-understood phenomena. But what is this third effect? As Lynch and Futterman properly note, "Apart from the difficulty in assigning a diameter to any of the arcs shown in Fig. 4.1, because they do not appear circular and therefore do not have a constant radius, it's unclear which arc is being measured."

That seemed to be an unanswerable objection. But the drawing looked familiar to me, and I was finally able to place it. It resembled the drawing in Minnaert's book *The Nature of Light and Color in the Open Air*[5] that illustrated the case of reflected dewbows A dewbow is a rainbow formed by drops suspended on the hairs of blades of grass (or presumably by some other mechanism). Such drops are still very nearly spherical in shape, but they are located approximately in the plane of the ground, rather than being suspended in the air. What's more, the eye and mind clearly perceive that they are on the ground, so that when you try to account for the shape of the phenomenon, you don't perceive it as an arc of a circle, suspended in the air (the conic section generated by cutting the 42° cone with a plane perpendicular to its axis), but as a hyperbola (the conic section generated by the intersection of the cone with a plane parallel to, but not containing, the axis). What's interesting is that when Minnaert chose to represent the reflected dewbows, he drew them as if they really *were* hyperbolas, as if the dewbows formed real marks on the ground, and the hyperbolas represent what a viewer floating directly overhead would then see. Of course, such an overhead viewer would see nothing at all—only the viewer with his back to the sun would see a series of illuminated points 42° from the anti-solar point, which he would interpret (unless he made an effort to see the arcs as parts of a circle) as hyperbolas.

This is exactly what the illustrator for de Ulloa and Juan has done. If we assume that the same observer is looking at the arcs as at the Fogbow and the Glory, then he is looking down the mountainside at these arcs that would, to him, appear to lie at the same angle from the anti-solar point, but which look to us, the floating observer, like hyperbolas, or at least like a family of some sort of curve. That would also explain why de Ulloa and Juan talk about such apparently noncircular arcs as having set diameters. As seen by the observer, from his location, each curve *does* have a well-defined angular diameter.

The explanation seemed to be correct in too many ways—a rainbow seen by moonlight, it is well known, appears white.[6] If the dewbow, its secondary, and a reflected dewbow were seen on a downward slope by moonlight, they would appear as three white hyperbolas, just as illustrated. I wrote to *Applied Optics* and to Dr. Lynch. He replied that such an explanation, for all its apparent aptness, still has difficulties. The observed angular diameter of the observed phenomenon was 60°, not 84°, as a dewbow would have been. Moreover, the angular width of 5° de Ulloa and Juan report is too wide for a rainbow or dewbow. And secondary moonbows are normally too dim to be seen.

It was a long time before a better explanation occurred to me. The other two phenomena de Ulloa and Juan observed and reported were observed in fog banks—why not this one as well? If the observed phenomenon was a series of low-lying fogbows, then they would appear to be white even in sunlight, and also in moonlight. It would have a definite angular diameter and yet would appear hyperbolic in shape, matching the engraving. But in this case the angular diameter of the bow would be smaller—fogbows are very peculiar forms of rainbows, formed when the droplet size is approximately that of the wavelength of light. Typical fogbow diameters are about 60°. Moreover, the angular width of a fogbow is wider than that of a conventional rainbow—about 5°, just as de Ulloa and Juan report for their arcs.

There is surprisingly little work on fogbows in the literature. One very useful report appears in the *London, Edinburgh, and Dublin Philosophical Magazine and Journal of Science* in 1890.[7] In this, James C. M'Connel reports on the theory and observations of fogbows. He cites a series of observations made at the observatory atop the mountain of Ben Nevis between May 1886 and October 1887. The angular widths and diameters were measured with a special instrument devised there, called the *stephanome*.[8]

Double fogbows were reported on November 12, 1886, December 26, 1886, February 13, 1887, July 21, and October 15, but no measurements were taken. On December 30, 1886, a double fogbow was carefully measured. The inner bow had an inner angular radius of 32° 20′ and an outer angular radius of 34° 44′. The outer fogbow, which was quite noticeable, had an inner radius of 36° 36′ and an outer of 41° 22′, giving it an angular width of almost 5°. They also reported lunar fogbows on at least five occasions.

M'Connel noted that double fogbows are a very common observation, far more common than one would naively expect. "Out of eighteen bows no fewer than ten were double," he wrote. The reason for this seems to be that fogbows, with the greater uniformity of droplet size and the smaller separation between peaks at different wavelengths, don't generally "average out" the first supernumerary bow the way larger raindrops do.[9]

There are two possibilities for the third bow—it might be the secondary fogbow, caused by two internal reflections rather than one. A secondary fogbow, like a secondary rainbow, is much dimmer than a primary, and especially so for a lunar bow, so this seems a very unlikely candidate.

The other possibility is that this is a reflected fogbow. This sort of bow, like a reflected rainbow, is due to light reflected from the fog droplets that comes not from the moon itself, but from the reflection of the moon behind the observer. Such a bow has the same angular radius as the original bow, but a different center. Because of this, the reflected bow deceptively appears to have a different radius. When only a portion of the bow is visible, in fact, it is often interpreted as a straight segment. Segments of such reflected rainbows are responsible for reports of "rainbow pillars."[10]

The reflected bow can be quite bright, almost as bright as the normal bow, if the reflection is itself bright and nearly specular. A reflection from ice or a plain of snow near glancing incidence would fit this requirement nicely and is not unlikely atop a 3,500 meter mountain at a temperature that might be near freezing.[11]

I suggest, then, that the unidentified phenomenon reported by de Ulloa and Juan is, in fact, a lunar fogbow, along with a primary reflected lunar fogbow. The bows are observed on a snow or ice-covered slope, from which a very low-lying fog is being emitted, so that the bow appears to be lying on the mountainside. The mind therefore interprets them as ground-hugging hyperbolas instead of interpreting them as circles suspended in the fog. The bows would not actually meet at the "closest" point, but they might easily give that impression. One of the brightest bows could have an angular diameter of 60° and a width of 5°, as M'Connell reports.

Notes

1. The entry on de Ulloa in *Appleton's Cyclopedia* (1887–1889) is online at http://www.famousamericans.net/antoniodeulloa/, accessed April 21, 2012.

2. Florida had been under the control of the British since the Seven Years' War (1756–1763). In any event, the Spanish recaptured portions of it, beginning with Pensacola in 1781, and regained control of the peninsula in 1784 as a result of the treaty ending the American Revolution.

3. Biographies of de Ulloa appear in the *Dictionary of American Biography* (1964), Vol. 10, pp. 107–108; *American National Biography* (1999), Vol. 22, pp. 95–96; and the *Dictionary of Scientific Biography* (1970), Vol. 13, pp. 530–531. See also references cited in these.

4. David K. Lynch and Susan N. Futterman, "Ulloa's Observations of the Glory, Fogbow, and an Unidentified Phenomenon," *Applied Optics* 30 (24): 3538–3541 (August 20, 1991).

5. M. Minnaert, *The Nature of Light and Color in the Open Air* (1954)

6. On Lunar Rainbows, see Jearl D. Walker, *The Flying Circus of Physics* (1975/1977), section 5.37, and the references therein.

7. James C. M'Connell, *London*, "The Theory of Fog-Bows" in *Edinburgh, and Dublin Philosophical Magazine and Journal of Science*, fifth series, 453–461 (June 1890).

8. The top of Ben Nevis was apparently a good place for making observations of the optical phenomena associated with fog. C. T. R. Wilson observed several glories while stationed there, and it was his desire to examine glories in the laboratory that led to his invention of the Wilson Cloud Chamber that came to play such an important role in particle physics. See James Burke, *Connections* (1978), 39, or see the second episode of his PBS series of the same name, "The Road to Alexandria."

9. For a good photograph of a fogbow, and a comparison to theory, see http://www.atoptics.co.uk/droplets/fogbow.htm, accessed April 21, 2012. Another picture is here: http://apod.nasa.gov/apod/ap061115.html, accessed April 21, 2012.

10. On Rainbow Pillars, see Walker, *The Flying Circus of Physics*, section 5.39 and references therein.

11. See Lynch and Futterman, "Ulloa's Observations of the Glory," 3541.

{5}

The Miracle of Saint Gascoigne

Saint Felix of Nola was said to have been born in the third century at Nola, near Naples in Italy. He distributed his inheritance among the poor, and was afterwards ordained a priest by Maximum, bishop of Nola. In 250 the emperor Decius began his persecution of the Christians, and they were forced to flee. Felix hid from his pursuers in a small cave or a small building. A spider came and wove a web over the opening, and Felix' pursuers, seeing the web, were certain that no one could be inside. And so St. Felix escaped.

The life of St. Felix[1] is supposed to be based on the poems and letters of Saint Paulinus, another Bishop of Nola who lived from about 354 to 431. But the same story of the miraculous escape is told of the Muslim prophet Muhammed while on his hijra.[2] The cave in which he is supposed to have hidden, the Jabal al Thur, is a point of interest visited by Muslim pilgrims during the hajj.[3] The same story is told about the future King David, hiding from King Saul.[4] It is also told of Yoritomo (1147–1199), one of the founders of the Shogunate in Japan.[5] The tale is told about others, as well, often in very general terms about an unidentified individual. The story of how the hero is protected by a spider weaving a web across the mouth of a cave or other hiding place, convincing his pursuers that no one lies within, is apparently simply too good not to be told. It illustrates different things to different people—the Power of Faith, the Power of God, that Even the Flimsiest Things have Value and Usefulness. It shows how something as light and insubstantial as a gossamer web can still restrain a powerful enemy.

That last point (which is the lesson King David was supposed to have learned) was true in optics as well. Spider webs used to be considered the best material for fine reticles.[6] "If only straight lines are desired nothing surpasses the fine web of the spider, which is easy to thread," read one old text on instrument making. There are several books and articles on the gathering, storage, and use of spider silk for crosshairs.[7] Users of spider silk have had to catch or raise spiders, keep them well and fed, and induce them to extrude silk. This is collected in various ways, such as winding it under tension on a fork. Threads can be obtained from orb webs or from cocoons as well. The spider silk can even be split to produce narrower webs. At the end of the nineteenth century, a variety of devices for "milking" numbers of spiders were developed by those interested in weaving spider silk. (A recent effort at obtaining and weaving spider silk, requiring the efforts of 70 people over four years, resulted in an 11 by 4 foot tapestry.[8]) Unfortunately, as spiders are territorial and cannibalistic, raising them for their webs is, unlike raising silkworms, an extremely labor-intensive and expensive operation. Fortunately for the world of scientific instruments, just at the time people were coming to realize this (and needing large quantities of material for crosshairs), C.V. Boys discovered the ductility of fused silica, and the ease with which it could be drawn into extremely fine fibers.[9] Crosshairs and reticles have, of course, been made of other threads and fibers and wires, or have been deposited, etched, or graven into transparent substrates.

The earliest knowledge of such crosshairs has been attributed to William Gascoigne of Leeds (1612–1644), who at first used literal hairs and textile threads. But about 1639 he made a wonderful discovery, very much by accident. An amateur astronomer, he was working on a Keplerian telescope, and when he resumed his work on it one day he discovered that there was something inside the barrel—a spider had built its web there overnight. When he looked through the telescope he found that not only were distant objects in focus, but so was the portion of the web that fell at the focal point of the lens, as well. Gascoigne saw in this discovery the miraculous Hand of God, as he wrote in a letter to mathematician William Oughtred[10]:

> This is an admirable secret which, as all other things, appeared when it pleased the All Disposer, at whose direction a spider's line drawn in an open case could first give me by its perfect apparition, when I was with two convex (lens)es trying experiments about the sun, the unexpected knowledge.

There isn't any evidence that Gascoigne tried to use spider silk crosshairs himself, although others did, later. But he had learned the secret of placing a reticle so that it could do the most good through what he was certain was the same sort of Divine Intervention that had saved the persecuted in the past.

Notes

1. http://www.newadvent.org/cathen/06033b.htm; http://en.wikipedia.org/wiki/Felix_of_Nola, accessed April 21, 2012; Donald Attwater, *The Penguin Dictionary of Saints* (1965), 128–129.

2. http://en.wikipedia.org/wiki/Cultural_depictions_of_spiders, accessed April 21, 2012.

3. http://wikitravel.org/en/Mecca, accessed April 21, 2012.

4. In *Midrash Alpha Beta Acheres d'Ben Sira*, 9; http://www.torah.org/learning/pirkei-avos/chapter4-3.html#, accessed April 21, 2012.

5. Richard Gordon Smith, *Ancient Tales and Folk-Lore of Japan* (1918), chapter 15; https://www.sacred-texts.com/shi/atfj/atfj17.htm, accessed April 21, 2012.

6. David P. Todd, "Notes on the Preparation of Reticles," *Phys. Rev.*, series I, 11 (1): 59 (1900). See also John W. Nystrom, *Improvement in Eye-Pieces for Telescopes*, US Patent No. 169,917 (1875).

7. Many are listed in J. Darius and P. K. Thomas, "Crosswires in a Guiding Eyepiece," *J. Phys. E—Sci. Instruments* 14: 761–765 (1981). See also R. W. Sinnott, "Crosshairs from Spider Web," *Sky and Telescope* 73 (1): 97 (1987).

8. *Wired*, September 23, 2008; http://www.wired.com/wiredscience/2009/09/spider-silk/, accessed April 21, 2012.

9. C. V. Boys, "On the Production, Properties, and Some Suggested Uses of the Finest Threads," *Phil. Mag.*, Series 6, 23 (145): 489–499 (June 1887).

10. Quoted in Silvio A. Bedini, "Along Came a Spider—Spinning Silk for Cross-Hairs," *The American Surveyor* 2 (2): 74–82 (March/April 2005).

{6}

Rays of the Sun

I sometimes think that we should spend at least a little time explaining everyday manifestations of physics to undergraduates, so that they can talk about phenomena that appear in everyday lives. Everyone studies *Electricity* and *Magnetism*, usually in their freshman year, but the introductory and even advanced courses don't discuss the most familiar form of magnetism—ferromagnetism—at all. It's brushed aside, hidden under a rug, because it doesn't fit neatly into the lesson plan. The freshman home on Spring Break has to excuse himself, saying that they haven't covered the topic yet. Most, in fact, never do get any real exposure to it.

Similarly, it has always seemed odd to me that probably the most familiar optical phenomenon never gets addressed. It's probably one of the first things we take notice of when we make our own survey of the optical world (fig. 6.1)—the rays of the sun.[1] The same image of a round solar disc surrounded by straight rays appears in Akhenaton's Egypt, 3,500 years ago.[2] The rays of the sun literally fill the sky in Pieter Breughel the Elder's 1570 print *Summer*.[3] But, in all my years in college studying and teaching optics, no one has ever addressed this phenomenon. Just what does cause solar rays?

"Scattering by dust," answered a colleague when I posed the question to him. But let's be clear about the phenomenon under discussion. I'm not referring to the shafts of sunlight breaking through clouds, creating relatively broad sunbeams which can be easily photographed.[4] Jearl D. Walker calls these the "Rays of Buddha,"[5] and they are certainly due to the sun's rays being scattered by dust or other matter. But the effect I'm referring to is very different in appearance and causation. It cannot be photographed and is evidently due to a limitation of the eye itself. Hermann von Helmholtz's book on Physiological Optics is almost a century old, but it is still one of the most useful compendia on the topic. In the English translation, a description of this effect is given in the section on monochromatic aberrations:

> With very intense light. . . . as when the pinhole is illuminated by direct sunlight, the points of the star seem to merge together, while all around it an immense number of exceedingly fine, brilliantly colored, radiating lines form a sort of corona of much greater extent. The name *hair corona* will be used to distinguish this phenomenon from the star-shaped blurred image.[6]

FIGURE 6.1 *Child's drawing of the sun, showing the characteristic rays.*

The rays also appear around automobile headlights and around the filaments of flashlight bulbs. Walker mentions this as well.[7] He explains the effect as seen in photographs as being due to diffraction at the edge of the aperture. He cites J. R. Meyer-Arendt,[8] which states that the effect in the human eye is due to diffraction at the irregular edge of the human pupil. Meyer-Arendt, in turn, gives no citation at all.

Others hold that the fine lines of the solar rays are due to the radiate structure of the lens itself.[9]

Which physical effect is responsible for these rays? Or is it due to both?

The diffraction from the edge of the pupil is one that, I must admit, occurred independently to me. I'm still not sure of the structure of the inner edge of the iris, but it seems likely to me that it must be something like a bag or a sweatshirt hood closed with a drawstring. When it contracts, the inner edge, like the top of a drawstring bag, ought to have a set of very fine yet irregular serrations. I have no assurance that this mental picture is correct, and I have looked in vain for corroboration from journal articles and from highly enlarged photographs of pupils. Certainly our experience with ubiquitous laboratory leaf-spring irises leads us to expect that physical irises are not perfectly circular. The curved-edge polygonal shape of laboratory apertures will produce diffraction images with spikes on them. The topic of far-field diffraction from polygonal apertures has been well-treated in the literature.[10–15] Diffraction by serrated-edge circular apertures, arguably a closer match for our case, has been modeled as well[16–18].

Helmholtz himself suggested that the rays might be due to "the minute indentations in the edge of the pupil." He noted that if an irregular aperture smaller than the pupil of the eye is placed before the eye one sees different rays, and that these rotate with the pupil.[19] M. Minnaert,[20] in his classic book, noted that when the edge of the pupil is obscured by the eyelid, one sees longer rays in that direction, and that their direction is normal to the eyelid at that point.

All of this is encouraging, but misses the directness of proof. Why not simply substitute a circular aperture known to be nearly perfectly round? Perhaps Helmholtz had no good apertures and had to make do with an irregular one. In any event, such apertures are easily available today. I found a 600 micron circular pinhole in our lab and held it up in front of my eye while looking at the bright filament of a Maglite flashlight.

The effect was dramatic. Although a little hunting in x and y were necessary, I was able to find a location where the Helmholtz *hair corona* completely disappeared. The radial filaments of light were replaced by dark circles, resembling Airy discs. They probably were Airy discs, although the number of visible concentric circles seemed somewhat large for the classic Airy diffraction pattern, for which the maxima of successive rings ought to decrease rapidly.

Nevertheless, this simple experiment seemed to be the most direct confirmation of the pupil-edge hypothesis. I saw no more filamentary rays. Nor did any of the others I asked to try the experiment.

I'd still like to see a proper experimental proof of this, such as measuring the true shape of the pupillary aperture and using it to generate a diffraction pattern. But the experiments and papers listed give me confidence that almost all, if not truly all, of the "ray" effect is due to diffraction at the pupil's edge. Is there any significance to this?

I think that the appearance of these fine filamentary "hairy corona" rays has had a profound effect on the development of optics. I suspect that the image of rays extending outwards from a bright source suggested the idea of such light rays to early investigators of Optics—Claudius Ptolemy, Alhazen, Descartes, and others, and that this gave us Geometrical Optics (and later, Physical Optics).

The geometrical ray we are so familiar with from our introductory classes in optics is an abstraction. It can be approximated by the "pencil of light" one can obtain by letting sunlight in through a pinhole, but these could only be used for relatively short distances and are weak. Even in works by the ancient Greek philosophers, however, light was thought to move in straight lines (although the direction they imagined the light to be moving was often reversed from our understanding).

In reality, the ray represents the normal to the wavefront, and a more "natural" way to treat the propagation of light might be to propagate this wavefront through space (Interestingly, Titus Lucretius Caro comes very close to suggesting this. In his long poem *De Rerum Natura*, he talks of "films" thrown off by objects, which are received by the eyes. But he did not try to produce a mathematical theory of this, nor was he in a position to do so.) Even in works that have introduced me to the Huygens-Fresnel integral, however, the formulation has been in terms of rays from geometrical optics.

It's not easy to visualize an entire wavefront, but we can easily visualize the progression of a single ray. This is precisely what Rene Descartes did when he produced his theory of the rainbow. He imagined a great many rays forming the wavefront, then traced the actual path of regularly spaced representatives from this vast assemblage of rays, noting how the rays from different points along the front were transformed by passage through the raindrop at different distances from the center. It wasn't feasible to treat the entire wavefront at once, but the ray optics approach allowed him to treat it piecemeal and get the same result.

In a way, this is parallel to the way the structure of the brain is studied. The tissue of the brain consists of closely packed neurons, but that was not clear in the nineteenth century. Investigators wondered about the true structure of the brain, but were unable to untangle it, until Camillo Golgi was able to develop a stain for neurons that selectively affected only a few neurons, leaving the others uncolored.

Notes

1. http://www.nasa.gov/lb/audience/forkids/artsstories/AS_More_Sun_Drawings_by_NASAKids.html, accessed September 22, 2012.
2. http://www.historical-reproductions.com/e-18.html, accessed September 22, 2012.
3. http://rubens.anu.edu.au/htdocs/bytype/prints/brueghel/0001/110.JPG and http://rubens.anu.edu.au/htdocs/bytype/prints/brueghel/0001/111.JPG, accessed September 22, 2012.
4. http://www.capetownskies.com/sunbeams.htm, accessed September 22, 2012.
5. Jearl D. Walker, *The Flying Circus of Physics with Answers* (1975/1977), 149.
6. Hermann von Helmholtz, *Helmholtz's Treatise on Physiological Optics*, ed. James P. C. Southall (1962; reprint of OSA 1924 translation of third German edition), 129.
7. Walker, *The Flying Circus of Physics*, section 5.97 on page 137.
8. J. R. Meyer-Arendt, *Introduction to Classical and Modern Optics* (1972), 3.
9. See, for example, W. D. Zoethout, *Physiological Optics*, 3d ed. (1939), 95–96 and figure 66.
10. J. Komrska, "Fraunhofer Diffraction at Apertures in the Form of Regular Polygons I," *Optica Acta* 19 (10): 807–816 (1972).
11. J. Komrska, "Fraunhofer Diffraction at Apertures in the Form of Regular Polygons II," *Optica Acta* 20 (7): 549–563 (1973).
12. E. Hecht, "Symmetries in Fraunhofer Diffraction," *Amer. J. Phys.* 40 (4): 571–576 (1972).
13. Salvatore Ganci, "Simple Derivation of Formulas for Fraunhofer Diffraction at Polygonal Apertures from Maggi-Rubinowicz Transformation," *J. Opt. Soc. America* A *1* (5): 559–561 (1984).
14. Winifred Sillitto, "Fraunhofer Diffraction at Straight-Edged Apertures," *J. Opt. Soc. America* 69 (5): 765–770 (1979).
15. Richard C. Smith and James S. Marsh, "Diffraction Patterns of Simple Apertures," *J. Opt. Soc. America* 64 (6): 798–803 (1974).
16. B. K. Yap and S. D. Fantone, "Application of a Sunburst Aperture to Diffraction Suppression," *J. Opt. Soc. America* 64 (7): 978–982 (1974).

17. Nicholas George and G. M. Morris, "Diffraction by Serrated Apertures," *J. Opt. Soc. America* 70 (1): 6–17 (1980).

18. Y. Kim, H. Grebel, and D. L. Jaggard, *J. Opt. Soc. America* A 8 (1): 20–26 (1991).

19. Helmholtz, *Helmholtz's Treatise on Physiological Optics*, 195.

20. M. Minnaert, *The Nature of Light and Color in the Open Air* (1954), 94–95.

{7}

Roy G. Biv

When I was very young, my father taught me the mnemonic used to remember the colors of the rainbow and their order—the name *Roy G. Biv*, standing for *Red, Orange, Yellow, Green, Blue, Indigo, and Violet.*[1] The mnemonic is of long standing. I have found a reference to it from 1869.[2] It's simple, short, and easy to remember, but there's something a little troubling about it.

Why only seven colors? We can easily divide up the spectrum into a great many more colors. Even if we restrict ourselves to easily distinguishable colors, there are many more than seven. My old Crayola crayon box held 64 colors, and, even discounting obviously inappropriate entries and those which are simply lighter or darker shades, there are still variants with exotic names such as "turquoise" and "magenta," as well as the more pedantic "green blue" and "blue green."

Some popular explanations hold that the number is dictated by mysticism—there are Seven Colors just as there are Seven Days of the Week, Seven traditional astronomical bodies (Sun, Moon, and the five planets known to antiquity), Seven Deadly Sins, Seven Cardinal Virtues, Seven Archangels, Seven Wonders of the World, Seven Hills of Rome, and so forth. Some even attribute a belief in seven colors to ancient philosophers, such as Aristotle.[3]

William E. Gladstone, British statesman and Homeric scholar, wrote a three-volume *Studies on Homer and the Homeric Age* in 1858, in which he has a section on "Homer's Perceptions and Use of Colour," wherein he observes that there are very few color terms in the *Iliad* and the *Odyssey*, and that Homer uses them in ways we would think bizarre—calling the color of cows and the sea by the same name, for instance, or using an adjective that appears to mean "blue" to describe someone's hair. He concluded that[4] "the organ of colour and its impressions were but partially developed among the Greeks of the heroic age." This was no hasty, slapdash judgment—Gladstone was intimately familiar with his Homer, and his discussion of his use of color takes up 42 pages, filled with quotations and citations. It touched off a long argument about color and its perception in the ancient and modern world that is still going on.

In any event, certainly the description of color was different in ancient Greece, if nothing else. Aristotle claimed three colors in the rainbow—Red, Green, and Blue.[5] Other Greek philosophers credited the rainbow with four colors.

In later centuries, Roger Bacon (1214–1292) claimed five colors[6]—White, Blue, Red, Green, Black. Theodoric of Freibourg (?–c. 1311) thought he could distinguish four colors in rainbows, prisms, spiderwebs, and the like[7]—Red, Yellow, Green, and Purple. The artist Cennino Cennini (c. 1370–c. 1440), in his book on art, *Il Libro Dell-arte*, posits seven paint colors for artists—Black, Red, Yellow, Green as "natural" pigment colors, along with three "Artificial" colors of White, Blue, and Yellow.[8] Carl Boyer claims that rainbow theorist Franciscus Maurolycus (1494–1575) believed there were seven colors, corresponding to the seven internal reflections he imagined were responsible for the formation of a rainbow (although I suspect Boyer is mistaken here).[9]

From all of this, it should be clear that there is no ancient, long-standing tradition of seven colors, or of any fixed number of colors, let alone which ones they were. Our modern theory of color and the naming of the colors can be attributed to one individual—Sir Isaac Newton (1643–1727), who transformed our understanding of physics and mathematics as well. It's in his 1704 classic *Opticks* that he first enumerates the colors as seven, and names them: Red, Orange, Yellow, Green, Blue, Indigo, and Violet.

And yet, it seems as if even Newton wrestled with this question. In his earlier *Optics Lectures* of 1670–72, he speaks of only *five* colors—Red, Yellow, Green, Blue, and Violet. Furthermore, if you examine the *Opticks* closely, you see that he switches between asserting five colors (ignoring Orange and Indigo) and seven, sometimes within the space of a page. For instance, in Book 1, Part 1, Experiment 9 (page 55 in the 1952 Dover reprint, which all my notes will refer to), he describes the colors obtained by breaking up sunlight with a prism. He names Green, Yellow, Red, Blue, and Violet, but not Orange or Indigo. He does the same in Book 1, Part 2, Experiment 4 on page 121. On page 123, he mentions "White, Grey, Red, Yellow, Green, Blue, Violet Bodies." On the next page, he describes Red-making light, Yellow-making, Green-making, Blue-making, and Violet-making, but not Orange-making, or Indigo-making. But on page 126, we get a listing of colors that includes Orange and Indigo. On pages 182–183, he describes the way water attenuates light of different colors and remembers to cite Orange-making and Indigo-making rays at one point, but seems to lapse later, talking about light affecting the Blue-making and Violet-making rays while ignoring Indigo-making rays.

Most telling, I think, are his descriptions of the colors he sees when he places a glass convex lens against a glass flat. He observes the concentric colored rings that we today call "Newton's Rings" in his honor (although they had been described by Robert Hooke in 1664), which we explain as an interference effect. Newton, committed to the corpuscular theory of light, rather than the wave theory, cannot use constructive and destructive interference to explain these colors, but he does come tantalizingly close to wave theory, observing that the colors are associated with characteristic separations of the glass surfaces that *almost* seems to identify them as integral multiples of wavelengths, but just misses. In his descriptions of the colors, he first identifies only five—Violet, Blue, Green, Yellow, and Red.[10] Newton is here seeing the effects of Multiple Order White Light Interference, which, except near the first order, doesn't give you a true full spectrum. Note that, except for the omission of Blue (possibly absorbed into "Purple" with the Violet), these are the same colors Theodoric of Freibourg saw in his observations of prisms and spiderwebs. And here, perhaps, is the reason for that paucity of colors, and the curious omission of Orange—orange shows up in a simple spectrum from a prism or a modern diffraction grating, but it's absent from multiple order white light interference, such as the colors in a skim of oil on water. It's even frequently missing from rainbows, unless the drops are significantly larger than a micron. Newton chose five colors because that was what he saw in most of his explorations of light and color.

It took something important to get him to change his mind and switch to seven colors, as he did on p. 212, only a page after asserting there were five colors. Newton came up with an astounding idea—the succession of colors resembled the succession of notes in a musical scale, and he became convinced that the ratios between colors corresponded to the ratio between successive tones.

It's an intriguing idea, and one that is commonly invoked today. For his 1987 PBS miniseries *The Ring of Truth*, physicist Phillip Morrison explained the different kinds of electromagnetic radiation by comparing them to the notes on a piano, with frequencies varying for electromagnetic radiation as they do for the sound waves generated by the keys of the piano. To emphasize this, he had the keys of the first octave, from Middle C on up, colored with the successive colors of the spectrum, starting with Red for Middle C. It was a graphic way to illustrate the spectrum, showing how ultraviolet, X-rays, and Gamma rays lay at higher frequencies to the right, while infrared, radio waves, microwaves, and millimeter waves lay to the left, at lower frequencies. It's even more remarkable that Newton came up with this analogy while not himself believing that light actually consisted of waves.

In the Pythagorean system of tuning, the different notes on, say, a lyre, were obtained by keeping the tension constant and changing the length of the string. Start with a string of given length and call the note "C" or *do*.[11] Halve the length of the string and the resulting note is an octave higher than your starting note—the next "C" or the higher *do*. Make the string 8/9 as long as it originally was and you get "D" or *Re*. Five-sixths gives you *Mi*, 3/4 gives *Fa*, 2/3 gives *Sol*, 3/5 gives *La*, and 9/16 gives *Ti*.[12] Newton observed that the same ratios give us the colors, using his characteristic lengths he discovered from his study of Newton's Rings. Except that he had to modify his original thesis of five colors to now include seven colors, corresponding to the seven notes from one *Do* to the next. And thus Orange, which had been simply a mixture earlier (possibly because it did not appear in the Newton's rings), now was promoted to the status of a color. There also needed to be a color between Blue and Violet, to correspond with the note *La*. Most people, looking at that expanse, see only blue and violet and shades thereof, but Newton needed another color, and so he invoked *Indigo*. Newton set up an experiment in which he asked another person to mark the limits of the seven colors and took the agreement with his estimation of the location of Indigo with his own as evidence that it was a proper division.[13]

Newton's ideas of colors corresponding to musical notes could have gotten a confirmatory boost from the observation of Newton's Rings, supernumerary rainbows, and multiple color orders seen in oil films—just as one musical scale an octave higher succeeds the previous one, ad infinitum, he could see successive spectral colors, starting over again with red. The progression is very clear with modern diffraction gratings, but with Newton's Rings and oil films successive orders of spectra tend to get muddled together and the colors lose their identity, so that eventually you simply have alternating bands of turquoise blue and violet pink. The idea of repeating spectra of all colors would have to be suggested by only the first couple of orders.

Newton's spectral color scale agreed well with the musical scale in both position and spacing, but it could be argued that *any* division of the spectrum could be made to agree with the spectrum. If, instead of the diatonic scale he used, corresponding to the white keys of a single octave of the piano, he could have used the full chromatic scale, with all its sharps and flats, corresponding to the white and black keys. We would have had five more colors added to the scale for a total of twelve, with half-steps between Red and Orange—call them *Pumpkin*, perhaps, or *Coral*—and between Orange and Yellow—*Amber*, perhaps? The point is that wherever you put your lines of demarcation, there will be a color there—and Red, Green, Blue, and Violet are so very broad, no matter how your colors are laid out (equal wavelength, or width determined by prismatic dispersion, or width in a rainbow, or what have you), that just about any scheme will get each of them. The only iffy color is Yellow, which often occupies a relatively short region. Any color scheme that didn't have Yellow as a major color would be judged deficient—but, fortunately, Newton's seven-tone scale hits Yellow pretty squarely.

And what of *Indigo*? It entered the English language to describe the color imparted by the Indicoferra dye plant, which was used to dye the denim to make heavy-duty work pants, giving *Blue Jeans* their name. Faced with such a color, I'd unhesitatingly call it *blue*. Some have argued that Newton had a different idea of Blue in mind,[14] or that they can easily distinguish the subtle (to me) difference between Blue and Indigo. The evidence suggests otherwise. When the Radio Manufacturers Association adopted its Color Code[15] for resistors (eventually published as EIA-RS-279), it started with black and brown for the numbers 0 and 1, then went up the spectrum for 2 through 7, conspicuously neglecting "Indigo," before turning to Gray and White to complete the sequence as 8 and 9. One assumes that it would have been mistaken for Blue or Violet. Similarly, the current rash of "Rainbow" flags, standing for diversity, tolerance, gay rights, and the like, tend to have only six colors, rather than seven, again keeping Blue and Violet while eliminating Indigo. In a few cases the flags feel compelled to include seven colors, but when they do the seventh is not an Indigo that falls between Blue and Violet, but a most out-of-place light cyan.

The seven colors, despite popular assertions that they were numbered thus out of a sense of mysticism, appear not to have been, but I suspect that the same love of a mystical seven is helping retain the notion of the "Seven Colors of the Spectrum." That, and a desire to let Roy G. Biv have a pronounceable last name.

Notes

1. According to Wikipedia, a common British mnemonic is *Richard Of York Gave Battle In Vain*, supposedly referring to his loss at the Battle of Wakefield. Yorkshiremen, unwilling to mention the defeat, used *Rowntrees Of York Gave Best In Value*. Other mnemonics include *Read Out Your Good Book In Verse* and *Rinse Out Your Granny's Boots In Vinegar*. http://en.wikipedia.org/wiki/Roy_G._Biv, accessed April 4, 2012. I haven't found references to these alternates anywhere near as old as for "Roy G. Biv" itself.

2. "The Nature and Laws of Light," *The Southern Review* (Methodist Episcopal Church) 5 (9): 18–36 (January 1869). The reference on page 28 makes it clear that the mnemonic is well-established by the time of the article.

3. Apparently Roger Bacon credited him with such a belief. See Carl B. Boyer, *The Rainbow: From Myth to Mathematics* ([1959] 1987), 102. William E. Gladstone *Studies on Homer and the Homeric Age* (1858), 494, claims that Aristotle knew seven colors, but they were Black, Gold, White, Red, Violet, Green, and Blue.

4. Gladstone, *Studies on Homer and the Homeric Age*, 488.

5. Boyer, *The Rainbow*, 48. Boyer also notes that Aristophanes also claims the rainbow to have three colors—but they're Purple, Green, and Yellow.

6. Boyer, *The Rainbow*, 102.

7. Ibid., 113.

8. http://www.colorbasics.com/HistoryOfColorScience/, accessed April 4, 2012 and http://en.wikipedia.org/wiki/Cennino_Cennini, accessed April 4, 2012.

9. Boyer, *The Rainbow*, 159.

10. Isaac Newton, *Opticks* (1704, although I use and refer to the Dover edition, published in 1952 and several times since), 198, and in the illustration on 209 and 211.

11. The vocalized scale is called *solfège*. Remember the song from *The Sound of Music*, or your old music classes—the notes are *Do, Re, Mi, Fa, Sol, La, Ti, Do*. Newton uses the same names in his *Opticks*.

12. There are different ratios that have been used for some of these notes, but they're very close to these.

13. Wikipedia article on *Indigo* at http://en.wikipedia.org/wiki/Indigo, accessed April 5, 2012, citing David Brewster's *Memoirs of the Life, Writings, and Discoveries of Sir Isaac Newton* (1855), Vol. 1, p. 508. http://books.google.co.uk/books?id=4hY6AAAAcAAJ&pg=PA408&dq=For+some+years+past,+the+prismatic+colours+being+in+a+well+darkene

d+room&hl=en&ei=9h3QTcGhL4r_-gaPrMTrCQ&sa=X&oi=book_result&ct=result&resnum=2&ved=0CD4Q6AEwAQ#v=onepage&q=For%20some%20years%20past%2C%20the%20prismatic%20colours%20being%20in%20a%20well%20darkened%20room&f=false, accessed April 5, 2012.

14. See, for instance, R. A. Houstoin, "Newton and the Colours of the Spectrum," *Science Progress* (October 1917) or his *Light and Colour* (1923) or David Topper, "Newton on the Colours in the Spectrum," *Studies in History and Philosophy of Science A* ***21*** (2): 268–279 (June 1990).

15. http://en.wikipedia.org/wiki/Resistor_color_code, accessed April 8, 2012.

{8}

Georg Christoph Lichtenberg

Everyone is a genius at least once a year. The real geniuses simply have their bright ideas closer together.

—G. C. LICHTENBERG

Georg Christoph Lichtenberg (1742–1799) was Professor at Göttingen, specializing in mathematics, astronomy, and natural philosophy (which we would call "physics"). He has been called "the first of the great German experimental physicists." He was also a noted wit. His book of aphorisms, *The Waste Book*,[1] is still in print in its English translation. He commented at length on the graphic works of William Hogarth. And he made extraordinarily bad puns.[2, 3]

Lichtenberg was born on July 1, 1742, in Oberramstadt, near Darmstadt in the state of Hesse in Germany, the 17th (!) child of Johann Conrad Lichtenberg, a noted theologian and pastor. The elder Lichtenberg was also known for his work in architecture, poetry, mathematics, and natural philosophy, and Georg seems to have followed in his footsteps.

He was short and had a large hump on his back, both of which tended to make him shy in large company and an outsider. It is perhaps not surprising that he had an active literary life.

After studying at the gymnasium at Darmstadt, Lichtenberg attended the University at Göttingen from 1763 to 1766. During this time he also made scientific excursions to central and northern Germany, and to Denmark. One object of these trips was to prepare astronomical charts for the sky as seen from the latitudes of the German estates of George III of England. Lichtenberg traveled to England in 1769, and again in 1774, carrying gifts for George III. While in London, Lichtenberg visited the theater frequently, collected artwork, and made general observations on city life. He published his observations on the English people in *Briefe aus England* ("Letters from England") in 1776–78.

He was made Extraordinary Professor at Göttingen in 1769, and Ordinary Professor in 1775, a post he held until his death. He was widely known as a witty and engaging lecturer, and he taught and performed research in geophysics, meteorology, chemistry, astronomy, mathematics, and physics. He is probably best known today for his work in electricity. He is believed to have installed the first lightning rod in Göttingen. In 1777 he was experimenting with discharging electricity into plates and found that if he placed a very sharp needle perpendicular to a nonconducting plate of resin, ebonite, or glass and discharged a Leyden jar through the needle into the plate it created distinct patterns. The patterns could be made visible by sifting fine grains of sulfur and red lead over the plate. These patterns, called *Lichtenberg figures*, are often of great beauty. They seem to resemble branching trees or river deltas, and efficiently cover the plate with a nonintersecting fractal network. They are still made today, although the materials and techniques have changed.

But what endears him the most to me are his commentaries upon people and on Hogarth's engravings. Not only is he a keen observer and expert at seeing tiny revelatory details in art, but he describes this in unfettered prose filled with his quirky humor. "It has not been my intention to make a footnote catalogue of all such plays on words," writes a modern commentator on Lichtenberg, "I resort only occasionally and out of self-defense... Now and again Lichtenberg is, for the modern sensibility, an exasperating punster."[4]

As an example of Lichtenberg's commentary, consider the following, from the Hogarth's series *Marriage à-la-Mode*:

> To compose a commentary on the steward's head, the meaning of his look, and the gesture of his hand would be the most inexcusable misuse of the alphabet. No letterpress in this world is cast for that. The two major divisions of the art of glossing would likewise perforce be doomed to failure in attempting such a text: both that whose aim is to aid comprehension, as well as the infinitely more learned one which aims at obfuscation. If I were to say: "Look, *this* is the very condition of His Grace's finances," and pointed at the figure of this household divinity, would anyone still ask, "Well, what *is* the condition of His Grace's finances?"[5]

But he could be surprisingly poetic and deep as well.

Most of the time, he sought to make observations and witticisms. *Non Cogitant, ergo non sunt*, he wrote, in contravention of Descartes's famous *Cogito ergo* sum.[6] Taking the word "nevergreen" from the poet Alexander Pope, he writes melancholically "The paths are bordered with Nevergreens"[7]

He could be risqué[8]:

> Her petticoat had very wide red and blue stripes and looked as if it were made of theater curtain. I'd have paid a lot for a front-row seat, but the curtain was never raised.

Alessandro Volta visited Lichtenberg in 1784. During supper, Lichtenberg asked Volta if he knew how to empty a glass of all of its air without the use of a vacuum pump. Volta said that he did not, at which Lichtenberg filled the glass with wine.

At this, Volta had to agree that the glass was empty of air. Lichtenberg then demonstrated how to let the air back in—he drank the wine. "This experiment seldom goes wrong if it is properly conducted," he remarked.[9]

Typical of his thinking and his pedagogical nature is an essay he wrote on "Consolations for the Unfortunates Born on the 29th of February," addressing the often-asked question of when someone born on February 29th ought to celebrate their birthday. Should they age only one-quarter as fast of the rest of us, or should they advance their birthday to the 1st of March or retard it to the 29th of February? Lichtenberg argues that neither of these is the correct solution.

Lichtenberg argued that each person's moment of birth is marked every year by the particular point where the sun stands in the ecliptic. Regardless of the calendar date, each moment of the year has a unique location specified by the solar longitude. The person ought to celebrate his or her birthday at the precise moment the sun reaches the same point it did the year of their birth.

An interesting thing happens—the calendar day corresponding to this moment in the year is not always the same, but drifts over of three or four days. More interesting still, this will be true of anyone who decides to celebrate their birthday using the solar longitude as a benchmark, even if it doesn't fall on February 29—our calendar is not so firmly fixed as we usually think it is, and our birthday will meander around over a space of a few days during the course of the years. "The man born on some other day [than February 29] who celebrates it according to the [calendar] date is often actually mistaken," Lichtenberg wrote, "But no one notices it."

To finish up this brief account of Lichtenberg, I note that his knowledge of optics[10] could still overcome his sense of artistic interpretation. In his commentary on the first of the series *Marriage à-la-Mode* he says, of the figure of the future Lord, "He is turned toward a looking-glass, but only because the glass hangs on that side where his betrothed is *not* sitting. He has in fact little or nothing to do with the mirror itself; the most he could glimpse in it would be a glint of silver brocade from his sumptuous sleeve. To assert (as does Mr. Ireland) that he could see himself or, indeed, even covertly eye his bride in the mirror, is a catoptrical impossibility." Yet modern commentators agree that the narcissistic young Lord is admiring himself in the mirror, whose placing might offend optics, but makes perfect viewing sense, just as the relationship between actor and mirror is frequently nonsensical in modern movies.

Notes

1. "The Waste Book" was Lichtenberg's own English name for the book, but I think it gives a false impression. As he himself wrote, these were the rough notes, comparable to those a shopkeeper would jot down, which were to be transferred later to a more formal Journal (Lichtenberg's term, again. I'd call it a Day Book), and, at greater intervals, into a formal ledger or log book ("Ledger at Double Entry," as Lichtenberg stiffly puts it). Lichtenberg felt that scholars ought to operate in the same way. Unless "Waste Book" was a real British term in use

at the time, he really means something like a Notebook or Scratch Pad. Lichtenberg's observations are generally too witty or insightful to be tossed away as "waste." See *The Lichtenberg Reader*, translated, edited, and introduced by Franz H. Mautner and Henry Hatfield (1959), 12.

2. http://www.quotationspage.com/quotes.php3?author=Georg+Christoph+Lichtenberg, accessed April 22, 2012.

3. *The Lichtenberg Reader* (1959).

4. *Hogarth on High Life*, from Georg Christoph Lichtenberg, *Commentaries*, translated and edited by Arthur S. Wensinger with W. B. Coley (1970).

5. Ibid., 39.

6. *The Lichtenberg Reader*, 87.

7. Ibid., 26.

8. Ibid., 50.

9. Carl Blinzer, *A Reasonable Rebel: Georg Christoph Lichtenberg*, trans. Bernard Smith (1960), 159.

10. Lichtenberg followed Newton and defended the corpuscular hypothesis of light. See Jürgen Teichmann, "Georg Christoph Lichtenberg: Experimental Physics from the Spirit of Aphorism," *Nuova Voltiana: Studies on Volta and his Times* 5: 91–103 (2000).

{9}

Hopkinson's Silk Handkerchief

Relatively few of the Founders of the United States of America are familiar to most Americans, which is a shame, since many of them were fascinating individuals with accomplishments in several fields. They have been completely eclipsed by the most famous of them—George Washington, Thomas Jefferson, Benjamin Franklin, John Adams, and a handful of others. Samuel Adams is probably best known to most people as a brewer because of the brand of beer named after him, which is ironic, as he may not have been a brewer at all, but merely a malt supplier.

One of the men who has fallen into obscurity is Francis Hopkinson, whom I myself should be aware of, if only because he was a delegate to the Continental Congress from my home state of New Jersey. A signer of the Declaration of Independence, Hopkinson was an accomplished musician, who composed the first American Opera and a songbook. He was a scientist who worked with Benjamin Franklin (he was the one who insisted upon using sharp points on electrodes to concentrate the "electric fluid") and was a member of Franklin's Philosophical Society of Philadelphia. He was an inventor (the Philosophical Society awarded him its Magellanic Prize Medal for one of his inventions) and a correspondent of Thomas Jefferson's. He was a poet and author of satirical works. He practiced as a lawyer and was eventually appointed judge of the US District Court for Eastern Pennsylvania. He was also a Trustee of the University of Philadelphia, which would one day become the University of Pennsylvania.

Hopkinson was sitting at his front door one evening in 1785, holding his silk handkerchief stretched between his hands and looking at a street lamp about 100 yards away, apparently out of curiosity and boredom. Out of this idle evening's recreation came some interesting and important optical physics. Hopkinson expected to see the threads of his handkerchief magnified, and decided that he wasn't disappointed, seeing the threads appear to be the size of "very coarse wires." For what followed, though, he had no explanation: "Although I moved the handkerchief to the right and left before my eyes, the dark bars did not move at all, but remained permanent before the eye.... To account for this phenomenon exceeds my skill in *optics*."

Hopkinson, unable to understand why the magnified threads apparently didn't move, decided to consult an associate who *did* have the necessary skill. Another member of his professional circle, also a member of the Philosophical Society and a Trustee of the College of Philadelphia, was America's foremost astronomer, David Rittenhouse.

Rittenhouse was a natural model maker from an early age, branching out into scientific instruments. He built two orreries, in exchange for which Rutgers University in New Jersey gave him a scholarship. A self-taught astronomer, he constructed his own telescope and made a careful measurement of the transit of Venus on June 3, 1769, and in 1781 observed Uranus. From 1779 to 1782 he was Professor of astronomy at the University of Philadelphia.

Rittenhouse examined Hopkinson's problem, writing up his results and delivering them to the Philosophical Society almost a year later.[1] "The experiment you mention," he began, "... is much more curious than one would at first imagine. For the object we see is not the web of the handkerchief magnified, but something very different." Rittenhouse recognized the effect as due to what Newton called the *inflection* of light, which today is called *diffraction*.

In order to study the effect more precisely, instead of using a normal handkerchief with its roughly uniform spacing, he constructed a model (as he had so often in his life), having a watchmaker put minute threads on pieces of brass wire, 106 to the inch, then laying down threads between the grooves thus created. Looking at a distance source, he saw four or five colored set of bars on either side of the central image. He rebuilt the device with threads 1/250 of an inch apart. The colored side bands were brighter and more distinctly colored. He used a measuring device from his telescope to measure the angles the different colors made and noted that the angles the sets of the same colors made with the perpendicular to the surface were integral multiples of the angle the innermost one made with the normal.

> I was surprised to find that the red rays are more bent out of their first direction, and the blue rays less; as if the hairs acted with more force on the red than the blue rays, contrary to what happens by refraction.... It is, however, consonant to what Sir Isaac Newton observes with respect to the fringes that border the shadows of hairs and other bodies.

He goes on to say that

> By pursuing these experiments it is probable that new and interesting discoveries may be made, respecting the properties of this wonderful substance, light, which animates all nature in the eyes of man, and perhaps above all things disposes him to acknowledge the Creator's bounty. But want of leisure obliges me to quit the subject for the present.

Unfortunately, he never found the leisure to take it up again. He was Treasurer of Pennsylvania until 1787, served in several offices at the Philosophical Society, becoming President in 1791, and became director of the US Mint from 1792 until shortly before his death.

If he'd had the opportunity, he could have pursued his experiments in diffraction and perhaps been the first to definitively measure the wavelengths of the colors of light. Because his pocket handkerchief model was the first consciously constructed diffraction grating, and the several bars of colors at multiple angles was clearly multiple-order diffraction. Thomas D. Cope estimated that Rittenhouse's measurements give about 620 nm for the red light and 460 nm for the blue.[2]

Of course, connecting the measurements with wavelengths requires a philosophical leap to the wave model of light. Whether Rittenhouse would have made this leap is now impossible to answer. Newton himself, in measuring the sizes of the colored bands associated with what we now call *Newton's Rings,* had associated characteristic lengths with the colors, but he was committed to the corpuscular theory of light and did not recognize these as wavelengths. It is interesting that, although Christiaan Huygens and Robert Hooke had championed the wave theory of light, they did not ask what the wavelength of the light might be, or its significance. It remained for Thomas Young, the indefatigable defender of the wave theory, to finally obtain his own grating and measure, for the first time, the wavelengths of the different colors in 1802.[3]Young didn't make his own diffraction grating, as Rittenhouse did. In fact, he didn't have to make anything at all—he used "commercial off the shelf" technology, in the form of finely graven reticules made by a Mr. Coventry of Southwark, England.

There's an interesting coda to the story. Francis Hopkinson went on to become a member of the Navy Board that wrote the Flag Act of 1777. In later years, Hopkinson claimed himself to have designed the flag of the United States. This has been disputed, but Hopkinson definitely had input into the design of the flag. This consisted of alternating red and white horizontal stripes, and an upper-left canton of white stars in a blue field, representing "a new constellation"—the United States. But it has been suggested that the Constellation also stood for that premier American astronomer, measurer of the transit of Venus and almost of the wavelength of light, David Rittenhouse.[4]

Notes

1. "An Optical Problem, Proposed by Mr. Hopkinson, and Solved by Mr. Rittenhouse," *Trans. Amer. Phil. Soc.* 2: 201–206 (1786).

2. Thomas D. Cope, "The Rittenhouse Diffraction Grating," *J. Franklin Institute* 214 (1): 99–104 (1932).

3. "Second Bakerian Lecture: On the Theory of Light and Colours," *Phil. Trans. Royal Soc. London* 92: 12–48 (1802).

4. Wikipedia articles on Francis Hopkinson and David Rittenhouse, http://en.wikipedia.org/wiki/Francis_Hopkinson and http://en.wikipedia.org/wiki/David_Rittenhouse, accessed August 22, 2010.

{10}

First Light

THOMAS MELVILL AND THE BEGINNINGS OF SPECTROSCOPY

On the evening of May 30, 1832, Évariste Galois sat up writing out mathematical formulas. He was only 20 years old, and he was to engage in a duel the next morning, which he was fated to lose. Fearing that he was to die, legend tells us, he frantically tried to set down all his revolutionary theories on paper before the death he was sure awaited him, scribbling "There is not Time!" on the margin of his paper.

It's a dramatic story, but when Tony Rothman researched the story—he intended to write a play based on the incident—he found that it wasn't true.[1] Galois did die in the duel, and did set down some work on the night before, but the frantic attempt to set down the remainder of his work for posterity with woeful asides did not happen. The situation, in fact, is dramatic and tragic enough without it. Galois did make contributions to Group Theory, Abstract Algebra, and other areas, and he would probably have gone on to make other contributions had his life not been cut so short. And so he goes to join the ranks of promising scientists who died too early, such as John Goodricke, the deaf-mute astronomer who, among other things, discovered and measured several variable stars, and died of pneumonia (contracted while observing stars on a dark night) at the age of 21.[2] Or of Thomas Melvill, who died at the age of 27 in 1753, less than two years after reading his paper "Observations on Light and Color" to the Medical Society of Edinburgh.[3, 4]

The credit for starting the field of Spectroscopy is usually given to William Wollaston in 1802. He observed the dark bands in the solar spectrum, which Fraunhofer later carefully measured. It seems a somewhat inverse method of founding a science more usually associated with light *emission*. Indeed, the field of Flame Spectroscopy is held to originate with the work of John Herschel and William Fox Talbot in the late 1820s, and it was sometime later that it was noticed that the dark solar bands corresponded to certain flame emission bands. But tucked away in a footnote is the observation that "some work" on flame spectra had been done in the eighteenth century by Melvill.[5] I have a love of beginnings and had never heard of Melvill. Who was this unsung founder of the science of Spectroscopy, and what had he actually done?

Thomas Melvill was a Scottish natural philosopher who attended Glasgow University. It 1752 he delivered a talk to the Medical Society of Edinburgh on two nights, a month apart—January 3 and February 7. In them, he addressed a number of topics on the nature of light and color, building upon the work of Newton. He examined and rejected the wave theory of light, agreeing with Newton's corpuscular theory. He then remarked upon the extremely small size and rarefied nature of these articles, which never seem to directly interact. He talked at length about the absorption of light by solid bodies and the attendant heating, observed the way that light is reflected from drops of water on certain plant leaves indicates that they are not fully in contact with those leaves,[6] then went off to discuss the appearance of colored objects under light of different colors. This naturally led to an explanation of how to generate lights of different colors in the first place. He took as his baseline "white" the light of an alcohol lamp. He then observed that "Bodies of all the principal colours, *viz.*, red, yellow, green and blue, are very little altered when seen by the light of burning spirits: but, if salts be continually mixed with them during the burning, different changes ensue."

I have not been able to find the earliest references to the use of foreign material to add color to flames. Certainly it was known by the nineteenth century, when packets of powder were sold that would produce different hues from a fireplace flame. You can find direction today on Internet sites for how to produce such effects. I have little doubt—but no documentation, I'm afraid—that such tricks have been known for a very long time. In the course of tens of thousands of years of throwing things onto fires—for a long time the only artificial source of light and heat—people must have learned of many substances that could be placed on a fire to alter the color. I note, for instance, that beechwood ashes contain so much manganese that they were used in the Middle Ages to color glass.[7] Melvill does not say how he came by his knowledge of flame coloration. It may have been a well-known procedure in his time.

He remarked on how observing colored objects by these colored lights dramatically changes their appearance. But then he went on from this experiment in perception to something more basic:

> Having placed a paste-board with circular hole in it between my eye and the flame of the spirits, in order to diminish and circumscribe my object, I examined the constitution of these different lights with a prism...and found that, in the first case...when sal ammon[iac], alum, or potash fell into the spirits, all sorts of rays were emitted, but not in equal quantities; the yellow being vastly more copious than all the rest put together, and red more faint than the green and blue.

This is, so far as I am aware, the first time that anyone took light from a source other than the sun or an unadulterated flame, ran it through an aperture and subsequently through a prism to break it into its individual components.

Sal Ammoniac is Ammonium Chloride, NH_4Cl, and it lacks a metal ion that would give color to the flame (and ammonium ion would not do so). Alum is $KAl(SO_4)_2{}^*H_2O$, while Potash is potassium carbonate, K_2CO_3. None of these would be expected to produce a copious yellow light. Potassium does have yellow lines in its spectrum, but they don't predominate—the classic potassium-colored flame is lilac in color. What Melvill was seeing was certainly the intense sodium D line, undoubtedly due to the contaminating sodium that often occurs in potassium salts. The fact that red was fainter than green or blue is consistent with the flame spectrum of potassium salts, in which green and blue are stronger than the longer wavelength lines.

> In the light of spirits mixed with nitre or sea salt, I could still observe some blue, though excessively weak and diluted; with the latter, the green was equally faint; but, with the former, pretty copious. But, when either of these salts were used, could hardly see any vestige of the red at all. . . .
>
> The proportion in which the bright yellow exceeds the other colours in this light, is still more extraordinary than in the former; insomuch that the hole seen through the prism appears uniformly of this yellow, and as distinctly terminated as through a plain glass.

This would again be pretty much consistent with modern observations. Nitre is Potassium Nitrate, KNO_3, and Melvill's sample must have had relatively much more sodium in it than his other salts did to make the red seem so dim. Sea Salt, made by evaporating sea water, could be expected to have predominantly sodium ions (30 percent), with Manganese next in prevalence (3.7 percent), followed by Calcium (1.2 percent), and then Potassium (1.1 percent). It's not surprising that green and blue should be weaker in this case, where there is much less potassium and with Manganese not producing visible spectral lines, while Calcium has red predominant.

Furthermore, Melvill observed that this light had a significant difference from sunlight or the light from a plain alcohol flame:

> Because the hole appears through the prism quite circular and uniform in color; the bright yellow which prevails so much over the other colours, must be of one determined degree of refrangibility; and the transition from it to the fainter colour adjoining, not gradual, but immediate

The spectrum of sodium has a distinct and well-defined line. This would have been easier to observe if Melvill had used a narrow slit, rather than a circular hole (as Wollaston would do a century and a half later, and which enabled him to see those narrow dark lines in the solar spectrum). Had he done so, he might even have been able to observe that the bright yellow color actually resulted from two closely spaced lines. But it's important not to judge him by the standards of our own knowledge, gained from standing on the shoulders of giants. Melvill was groping through country utterly unknown to him or to anyone else, and his discovery that the spectrum was discrete, rather than continuous, was a major discovery that could have allowed for a conceptual leap.

Could have, but did not. Melvill's topics go in different directions after this, and within two years he was dead. It's possible, although not certain, that he might have revisited this topic and pressed onward, and measured the relative positions of lines, establishing the science of spectroscopy a century and a half earlier than it was founded.

But perhaps not. Claudius Ptolemy in the first century built experimental apparatus and performed careful measurements of the refraction of light through interfaces between air, glass, and water, producing results very nearly identical to modern measurements, as we see in the chapter on "Claudius Ptolemy's Law of Refraction." He came close to discovering Snell's Law, but didn't quite do so. He was distracted by a different mathematical formulation and settled on an incorrect model. Neither he nor his students revisited the phenomenon. Had they done so, they might have noticed that the formula didn't work for small angles (most modern accounts of Ptolemy's work incorrectly state that it was correct at small angles), and the science of Optics might have advanced much more rapidly than it did. In the long run, it doesn't matter. Several others discovered the law of refraction (even before Snel), and Wollaston, Herschel, Fox, Fraunhofer, and others eventually rediscovered the basics of spectroscopy that Thomas Melvill unearthed, but which were forgotten.

Notes

1. Tony Rothman's article on Galois's last hours appeared in *American Mathematical Monthly* 89: 84 (1982). It is available online at http://www.physics.princeton.edu/~trothman/galois.html, accessed September 22, 2012.

2. On John Goodricke, see Stephen R. Wilk, *Medusa: Solving the Mystery of the Gorgon* (2000), 111–115; and Michael Hoskin, "Goodricke, Piggott, and the Quest for Variable Stars," *Journal for the History of Astronomy* **10**: 23–41 (1979).

3. Melvill's work was printed in the relatively obscure *Essays and Observations, Physical and Literary*, No. IV. It was, fortunately, reprinted in the *Journal of the Royal Astronomical Society of Canada* 8: 231–272 (1914). Available online at http://adsabs.harvard.edu/full/1914JRASC...8..231M, accessed September 22, 2012.

4. Biographical material on Melvill appears in the *Dictionary of Scientific Bibliography*, ed. Charles Coulston Gillispie (1974), Vol. 9, pp. 266–267.

5. *Literally* in a footnote—it appears on page 2 of William McGucken, *Nineteenth Century Spectroscopy: Development of the Understanding of Spectra 1802–1897* (1969) at the bottom of the page, and nowhere else in that book.

6. Thereby anticipating later work on *heiligenschein*, see chapter 25 "Retroreflectors."

7. On the use of beechwood ashes for coloring glass, see the footnote by John G. Hawthorne and Cyril Stanley Smith on pp. 55–56 to Theophilus, *On Divers Arts* ([1963] 1979).

{ 11 }

Mediocrity and Illumination

> Now I go to become a ghost myself. I will stand in the shadows when you come here to this earth in your turn. And when you feel the dreadful bite of your failures—and hear the taunting of unachievable, uncaring God—I will whisper my name to you: 'Salieri: Patron Saint of Mediocrities!' And in the depth of your downcastness you can pray to me and I will forgive you.... Mediocrities everywhere—now and to come—I absolve you all. Amen!
>
> —ANTONIO SALIERI IN *AMADEUS* BY PETER SHAFFER[1]

Playwright Peter Shaffer never claimed that his play *Amadeus* was historically accurate. *Amadeus* was no more the true story of Wolfgang Amadeus Mozart and Antonio Salieri than his *The Royal Hunt of the Sun* had been the story of Francisco Pizarro and the Incan Emperor Atahuallpa, nor *Equus* the story of a British boy who blinded six horses. Shaffer took his inspiration from real incidents, but he wove these facts into allegories about the relationship between God and Man. *Amadeus* asks how it could be that a just and loving God bestowed his Gifts of Music upon an undeserving and ungrateful boor, while leaving the virtuous and devoted Salieri a mediocrity. And so, Salieri, slighted, as he thinks, by God, wages war upon Him in the person of Mozart.

Except, of course, that Salieri was not a musical mediocrity. He was Kapellmeister to Emperor Joseph II in Vienna for 36 years, was president of the Musical Society, taught an array of famous pupils (including Beethoven), and composed a great many Operas and pieces of Church Music. Since the play appeared, and especially since the appearance of Milos Forman's film version, there have been restagings of Salieri's operas and concerts of his music. New CDs of his works have been released in recent years, including Cecelia Bartoli's *The Salieri Album*. Salieri was not mediocre—simply overlooked.

In the world of Physics, I have often felt as if Thomas Young (1773–1829) was the Patron Saint of Mediocrity. Not, I hasten to point out, because he or his works were mediocre. Young has an enviable track record, championing the Wave Theory of Light, doing important work on the Accommodation of the Eye, expounding on Elasticity, writing numerous articles for the new *Encyclopedia Britannica*, helping found Egyptology, and being a practicing physician on top of all that. But in two of his most famous endeavors, Young came out second-best. His work on translating the Rosetta Stone has been overshadowed by Champollion's work, and his formulation explaining the Supernumerary Rainbow was incorrect.

In the first case, Young had no need to be ashamed. He was an expert on languages, speaking over a dozen of them. His was the first successful work on the Rosetta Stone, and he succeeded in translating the Demotic Script (which he called "Enchorial") and made the first inroads into Hieroglyphics. His work was published well before that of Champollion, who eventually succeeded in deciphering the Hieroglyphics completely, but who never acknowledged Young's prior achievements.[2]

As for the latter, that is one of the rarely told stories of physics. All too often in the history of science the work of scientists past is glossed over. They become figures like prairie dogs, invisible until and unless they surface to grant their name to a formula or a principal or a constant. I knew of Young because of his Modulus of Elasticity and his two-slit experiment and even his work on hieroglyphics.[3] I had to wonder about the sort of person whose claims to fame were in areas as widely scattered as Optics, the Mechanical Properties of Materials, and Egyptology.

Thomas Young started as a classically trained scholar, whose interests could go in any direction. Early on his interest in the Eye took him to study accommodation, and from there his interest in light led him to study that subject, and to challenge the then dominant *corpuscular* theory of light. Today, with the advantage of hindsight (and knowing the Einstein and the photoelectric effect wouldn't come around to bedevil the issue for another century), the Wave Nature of Light seems to be an obvious thing. But it was a live issue at the time, and many objected that no conceivable medium could support light rays yet offer no other physical manifestation, or that the wave theory could not explain the fact of extraordinary refraction in crystals such as Iceland Spar. Against these objections, Young struggled to produce convincing demonstrations and explanations. He looked for evidence of the Wave Nature of Light everywhere and found it in reflections from fine scratches, soap films and oil films, diffraction around edges,[4] and the supernumerary rainbow.

The Supernumerary Rainbow is the name given to the "extra" rainbow arcs located within the curve of the Primary Rainbow, after Red, Orange, Yellow, Green, Blue, Indigo, and Violet. Under the right conditions, one may often see extra bands within these, alternating in purple-pink and blue-green. Newton did not mention these, and may have been unaware of them, although Edme Marriotte had described them in the seventeenth century. Others tried, without success, to explain them.[5]

To the modern opticist, the alternating green-blue and purple-pink coloring suggests multiple order white light interference, characteristic of oil films and all but the thinnest soap bubbles. Young, too, may have made the analogy with these effects, and surmised that the effect was due to the superposition of light waves traveling different paths and *interfering* with each other (a term he was the first to use).[6]

In the cases of thin films and plates, Young explained the variation in color as being due to interference between reflections from the front and rear surface of the plate or film. If light was a wave, as he surmised, then the two reflections would be of opposite sign and cancel each other when the optical path length between the two reflections was equal to half a wave. This effect was inexplicable using the corpuscular theory of light, but developed naturally from the wave theory of light.

Young explained the supernumerary rainbow in the same way. The rainbow occurs, as Rene Descartes showed, where rays exiting the spherical raindrop form the smallest angle relative to the line between the drop and the sunrays passing through the center of the drop are undeviated, but the angle of deviation increases as the distance between the point where the ray enters the drop and the center of the drop increases, until the point where the ray strikes the drop at 86 percent of the drop radius. The angle between the outgoing ray and the line to the sun is then about 42°, the smallest it will get. If the ray strikes the drop farther from the center than this, the angle begins to increase again. Descartes's view was that the rainbow appeared where it did because, although rays are emitted from the drop in almost all directions, the bulk of them are directed near this minimum deviation angle. That explains the one bright peak, but not the supernumeraries.

Young's explanation was that pairs of rays, one impacting on either side of that critical ray, would exit from the drop along the same angle relative to the sun, but with different path lengths. Along directions where the path length difference amounted to an integral number of waves, these rays would reinforce each other, and you'd get a bright supernumerary band. There would be a great many such subsidiary bands, all located inside the main rainbow and weaker than it, just as observed. What's more, such supernumerary rainbows would only be visible when the drops were very nearly the same size, since only then would there be a large effect from many drops adding together and giving maxima along the same direction. If the sizes of the drops varied, then the path length differences along any given direction would vary as well, and there would be no cumulative effect. The competing sets of Supernumeraries from each different size of drop would "wash out." That explained why supernumerary rainbows were not always seen. You needed to not only have a collection of drops in the right place, as for the major rainbow, but you also had to have a restricted range of droplet sizes.

Young's explanation seemed to cover all the observed facts. I was therefore surprised when I saw in Charles Boyer's book a reproduction of a graph from George Biddell Airy's important paper "On the Intensity of Light in the neighbourhood of a Caustic" from 1849.[7] It showed three plots of rainbow intensity vs. emergence angle. One represented Descartes's rainbow, with a smooth curve starting out low and gradually increasing toward infinity at the Cartesian Rainbow Angle. Another showed Airy's physical Optics calculation, with a series of oscillations—the Supernumerary Rainbows that Young sought to explain. The peak intensity is near, but not at, the Cartesian Rainbow angle, and the intensity does not stop at the Cartesian angle, but gradually tails off outside it. The third plot, however, showed the predicted intensity from Young's theory. It has the Supernumerary peaks, but they are out of phase with those of the Airy function by about 1/8 of a cycle. The intensity reaches up toward infinity at the Cartesian rainbow angle, and it ceases abruptly at the Cartesian angle. It appears, in other words, as if Young was wrong. He had the right idea going in, but failed in the execution, just as with the Egyptian Hieroglyphics. As if Young were playing Salieri to Airy's Mozart.

How could Young have miscalculated so badly? Having found a satisfactory explanation in general, how could he fail in execution? I obtained a copy of Airy's original paper. Airy is careful to not attribute that graph to Thomas Young. He never refers to it as "Young's Theory," but as "the imperfect theory."

It's clear that Airy is criticizing the direct and uncritical application of Young's explanation to the calculation of the intensity. If you assume that you only need consider the pair of rays—one from each side of the Cartesian ray—that emerge parallel from the raindrop in calculating the intensity, then you will get that "imperfect theory," with its sharp cutoff and phase error. The true calculation must take into account that every point on the wavefront acts as a nucleus for wave formation, as in the Huygens-Fresnel theory, and so one must integrate the contributions from each point on the wavefront, not merely from the two points the geometrical optics predicts will contribute to that angle.

A little thought shows that the Imperfect Theory really is demonstrably erroneous in a number of ways. It is not a geometrical or a physical optics formulation, but an incomplete combination of the two. It assumes strictly geometrical wave propagation only in a direction normal to the wavefront, but assumes that the light is in the form of waves that interfere. It assumes waves of all wavelengths, down to infinitesimal, high-frequency waves. A real wave calculation must show the limitations imposed by the finite size of the waves. An infinitely high peak in the intensity at the Cartesian angle, and the sharp and complete cutoff beyond are impossible unless waves of all wavelengths, including those of vanishingly small size are included. Yet the visible rainbow must be limited to wavelengths of visible light, as Young well knew.

Young never saw the graph of intensity vs. emergence angle. If he did, the unphysical nature would surely have struck him with force. He did make the mistake of assuming geometrical propagation in his verbal description of the interference phenomena, it is true, but his point was made. In all other examples—the single-slit and double-slit experiments, Newton's Rings, Interference from Parallel Scratches (very nearly diffraction gratings)—all properly take into account the Huygens-Fresnel nature of wave propagation. In these cases the wavefronts are all planar or spherical (or cylindrical). Young ran afoul of the Huygens-Fresnel principle in the case of the rainbow because the wavefront is *not* a simple planar or spherical wave in this case—it is a cubic (to first order, as Airy calculated it), with a pronounced cusp. He assumed that the same explanation of path length differences along parallel rays would suffice as it had before. This time, however, it did not.

Young was interested in providing evidence for the wave nature of light, and that was foremost in his mind when he proposed his explanation of supernumeraries. Having found an excellent example from nature to bolster his case, along with a simple, intuitively grasped explanation, he sought no further to explain his observations. He was still in the thick of defending his case for the mere existence of the wave theory. Airy was able to perform his calculations 35 years later, when the arguments had cooled down and the wave theory was in the ascendant.

So Young's theory of the rainbow did contain a flaw which he missed in the more immediate fight over the Nature of Light. Airy himself acknowledged Young's achievements and genius: "I well knew that in writing on *any* physical subject it is but ordinary prudence to look at [Young's work] first."[8] Young was no mediocrity. To cite his most recent biographer, Andrew Robinson, "Young made mediocrities uneasy, and to cover their unease, they belittled and ridiculed him."[9]

Notes

1. Peter Shaffer, *Amadeus* (1981).

2. If there is any comfort in the afterlife, it is that most people I've talked to not only don't know of Young's contributions, they had never heard of Champollion, either. The rivals are now united in relative popular obscurity.

3. His first lecture on the topic, surprisingly, did *not* feature his famous two-slit experiment, but an observation of light diffracting around a thin obstruction. The two-slit experiment was described six years later in print. It has recently been suggested that Young never even performed the experiment, but used it simply as a *Gedankenexperiment* to explain the principal.

4. There are three book-length biographies of Thomas Young (and countless briefer ones). The most recent is Andrew Robinson, *The Last Man Who Knew Everything* (2006); The others are Alexander Wood and Frank Oldham, *Thomas Young, Natural Philosopher* (1934); and George Peacock, D.D., *Life of Thomas Young, M.D., F.R.S. &c.* (1855). Peacock was an acquaintance.

5. An excellent and very accessible treatment of both the geometrical optics rainbow and the Airy Theory is in R. A. R. Tricker, *Introduction to Meteorological Optics* (1970).

6. Young's own work is available in *Miscellaneous Works of Thomas Young, M.D., F.R.S., &c.*, edited by George Peacock (1855), but reprinted in 1972 by Johnson Reprint Co. of New York & London. The critical paper on interference, "Experiments and Calculations Relative to Physical Optics" appears there, and in the *Phil. Trans. Of the Royal Soc. London* 94: 1–16 (1804).

7. George Biddell Airy's paper appeared in the *Trans. Cambridge Phil. Soc.* 6: 397–403 (1838), with Addendum in 7: 595–600 (1849). Both are reprinted in two different editions of the SPIE Milestones series, but without graphs or illustrations. Craig F. Bohren, ed., *Selected Papers on*

Scattering in the Atmosphere MS-7 (1989), 329–357; and Philip L. Marston, ed., *Selected Papers on Geometrical Aspects of Scattering MS-89* (1994); the graph appears in Carl B. Boyer, *The Rainbow: from Myth to Mathematics* (1959). It also appears (without the Descartes result) in M. V. Berry, "Exuberant Interference," *Phil. Trans. Royal Soc. London A* 380: 1023–1037 (2002).

8. Robinson, *The Last Man Who Knew Everything*, 186.

9. Ibid., 63.

{ 12 }

Even If You Can't Draw A Straight Line...

... You Can Draw Your Family, Friends, Anything from REAL LIFE—Like An Artist.... Even if You CAN'T DRAW A Straight Line!

So read the advertisements on the back pages of my comic books, when I was a kid back in the 1960s. The picture on the ads showed a person seated at a drawing table, looking through some odd prismatic device down at a sheet of paper upon which he was... *drawing!* Dashed lines from his eyes proceeded down to the sketch on the table, but another line was shown reflected from the prism and going horizontally over to the scantily-clad girl who was the subject of the drawing. It seemed to suggest that the mysterious device being offered for sale somehow projected an image of the subject down onto the paper, so that all you had to do was trace the outlines to produce a photographically perfect Work of Art.[1]

I was never tempted to send in the $1.98 that the wonder device cost, although I did ponder how the device must work. I eventually worked out that I had the lines of sight a little mixed up—surely the line of sight went from the subject, through the prism, and up to the viewer's eye, not down to the paper. There was no projected image. That would surely have been washed out by the ambient light. I also realized that the quality of the image would also depend upon the skill of the sketcher, no matter what the ad copy said. Indeed, clumsiness in the sketching of these images is in part responsible for the invention of photography. But we're getting ahead of our story.

The odd-looking device, I eventually learned, was a *camera lucida*, sometimes colloquially known (in art classes) as a *Luci*. It was invented, some time between 1800 and 1806, by William Hyde Wollaston, whose name is a familiar one in optics. He was independently wealthy, a physician and a scientist whose interests ranged from chemistry (he discovered palladium and rhodium in 1803) to physics. He was, unfortunately, no artist, and he felt the lack. Scientific illustrations were becoming important in publications. A colleague, the Reverend Henry Hasted, recalled later in life that Wollaston and he were limited in their ability to sketch the geology they were both interested in. "We could only take the *outline* of the districts we were interested in, for neither of us could draw well, and we lamented our not being able to do so." Fortunately, Wollaston's wide-ranging mind was perpetually active, and he noticed that he saw a double reflection of his face in his mirror as he shaved one morning. One reflection came from the rear, silvered surface, but there was a dimmer reflection from the first glass surface. Hasted, "calling on him a few months afterward in town, . . . found him with a minute truncated and half-silvered prism fastened with sealing wax to a piece of wire. 'Look,' said he, 'Here is the very thing we wanted at the Lakes.' "[2]

In the title of the patent he received (No. 2993 in 1806) (fig. 12.1), Wollaston simply called his device a "Drawing Apparatus," but in the body of the text it is referred to as "An Instrument Whereby Any Person May Draw In Perspective."[3] He was soon calling it a *camera lucida*. The name was almost certainly chosen by analogy with the more familiar *camera obscura* (literally a *dark room*), in which a lens (or pinhole) in the wall of a darkened enclosed room or box will allow one to project an inverted image on a screen, on which it could be traced. Wollaston's device, by contrast, didn't require such an enclosure. It could be used out in the light, and so qualified, arguably, as a *light room*.

Wollaston included several variations of the device in his patent. In the simplest, a plain or partially silvered mirror is placed at an angle to the horizontal, so that if one looks directly downward through the glass one can both see a piece of paper placed beneath it, and also whatever the partial mirror reflects. A second mirror, this one a fully silvered mirror, is placed at a 45° angle to the first partial mirror. This will guarantee that the angle between the incoming line of sight and the line from the eye to the first element is 90°. Why not simply place a single half-mirrored surface at 45°? In part because the reflected image would seem to be upside down. You *could* draw an upside-down image, but it's much easier and more natural to sketch an erect image.

Wollaston realized that there were other possible configurations. One is to place the two mirrors so that there is an angle of 135° between them. This still puts the downward line of sight at a right angle to the line of sight to the object being sketched. But Wollaston realized that there was yet a simpler way to arrange the surfaces—instead of putting the two surfaces on separate pieces of glass that could become disoriented relative to each other, one could simply make the two reflecting surfaces two adjacent faces on a specially-shaped prism. Wollaston mandated an angle of 135° between these two faces, and a 90° angle at the opposite side of the quadrilateral prism. The other two angles must add up to 135°, and Wollaston stated that they could each be 67.5°, but that it was not required.

There is, in fact, a slight problem with Wollaston's drawing of this situation. He shows the line of sight through the eye going through that first surface at normal incidence. If the nearest angle is 67.5°, then the angle of incidence at that first interface will also be 67.5°, and the light will be totally internally reflected unless the refractive index of the prism is an impossible 1.08. The prism cannot be used unless the angle is reduced and the prism rotated so that the ray that strikes the

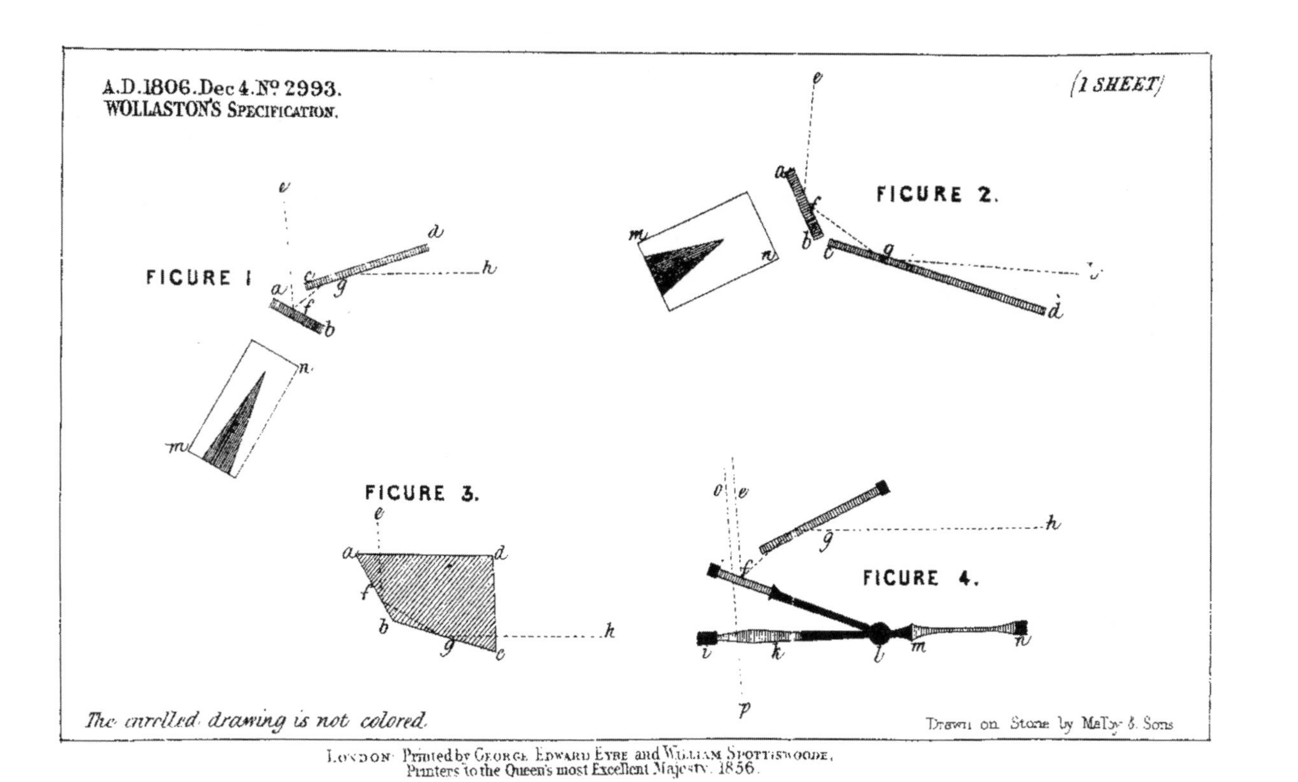

FIGURE 12.1 *Wollaston's Patent Drawing for the camera lucida.*

inner surface does so at an angle of less than 41.8° (if the refractive index of the prism is 1.5). One advantage of using the prism is that no silvering is required—the rays will strike the second surface at an angle greater than the critical angle, so it will be totally internally reflected. So long as the entering and exiting faces are at right angles, the incoming beam from the subject and the line of sight down to the paper will still be at right angles, despite any refraction the light makes upon entering or exiting the prism.[4]

Wollaston also added lenses in front of the mirrors or prisms, allowing "close-ups" of the subject, and an aperture in the brass mount of the device to force the eye to a single viewing point and to balance the light levels between the scene being viewed and the paper it is being drawn upon.

The *camera lucida* was instantly popular. It was a useful crutch for the nonartist. As Wollaston intended, it became a mainstay of scientific illustration, not only in geology but also as a means of preserving images seen through the microscope and the telescope. Some became enthusiastic *lucida* users. The scientist John Herschel was a fan and continued to make not only technical illustrations but also personal sketches for over 50 years (fig. 12.2). (An exhibition of his work, *Tracings of Light*, was on display in 1988-89 at the Baltimore Museum of Art and later published in a single volume.)

Not everyone who wished to use it found success, however. Some artistic skill was needed, even to translate the traced image into a pleasing drawing. As one writer observed in 1840, "Let not any one imagine that he can learn to draw, merely by purchasing a Camera Lucida; he might as soon learn music, by merely buying a fiddle." One who thus discovered his limitations was William Henry Fox Talbot, a

FIGURE 12.2 The Cape of Good Hope, *a camera lucida drawing by John Herschel.*

classicist, mathematician, and scientist. He was on his honeymoon in Italy trying to use the camera lucida, and recorded his frustration:

> [In] October, 1833, I was amusing myself on the lovely shores of the Lake of Como in Italy, taking sketches with a *camera lucida*, or rather, I should say, attempting to make them; but with the smallest possible amount of success.... After various fruitless attempts I laid aside the instrument and came to the conclusion that its use required a previous knowledge of drawing which unfortunately I did not possess.... It was during these thoughts that the idea occurred to me... how charming it would be if it were possible to cause these natural images to imprint themselves durably and remain fixed on the paper!

Talbot didn't merely wish for such images—he applied his knowledge of optics and chemistry to the problem, and became one of the pioneers of photography, inventing the negative/positive process that lay at the heart of most photography, developing photographic lenses (and in the process discovering the *Talbot Effect*), and going on to become, ironically, one of the founders of photography as an artistic medium.

Today, camera lucidas, both new and antique, are available through the Internet, as are instructions for constructing your own. They are used as aids to beginning artists and, as they were 200 years ago, as devices for hobbyists. Far from that 1962 price tag of $1.98, camera lucidas today sell for $200–$500 for a new one, and several thousand for antique originals. I wish I'd bought one when I first saw that ad.

I was surprised to learn that David Hockney and Charles Falco suggested that the *camera lucida* was employed by Renaissance artists, in addition to the *camera obscura* that features more prominently in their controversial hypothesis. I don't wish to start a discussion of this inflammatory topic[5] (which has already been discussed at considerable length by others), except to say that it seems unlikely that a Wollaston-type device was in use so early, and if it had been used there would certainly have been some mention of it. The *camera obscura*, at least, was certainly known to exist at that time, and is well attested to in the literature.[6]

Notes

1. The advertisement appeared in comics from American Comics Group in 1962.

2. For this, and for much of the material in this article, I owe a great debt to Larry J. Schaaf, *Tracings of Light: Sir John Herschel and the Camera Lucida* (The Friends of Photography, 1990). He gathers together much material on the history of the *camera lucida*. This is the book associated with the 1988–1989 exhibition of Herschel's *camera lucida* sketches exhibited at the Baltimore Museum of Art.

3. Wollaston's patent was published in *The Repertory of Arts, Manufactures, and Agriculture*, 2nd series, 10 (57): 161–164 (February 1807). It is available online at http://www.camera-lucida.com/images/Wollaston_s_Drawing_Apparatus_Patent.pdf, accessed September 22, 2012.

4. The ray paths shown on the website at http://painting.about.com/od/oldmastertechniques/ss/camera_lucida.htm, accessed September 22, 2012, do not correspond to the way

Wollaston used his prism. That path has a wasteful double reflection from the partially silvered mirror and does not have a right angle between the input and output beams.

5. David Hockney and Charles M. Falco, "Optical Insights into Renaissance Art," *Optics and Photonics News* 11 (7): 52–59 (July 2000); David G. Stork, "Optics and the Old Masters Revisited," *Optics and Photonics News* 15 (3): 30–37 (March 2004); Letter (Falco Response to Stork), *Optics and Photonics News* 15 (6): 7 (June 2004); "Scatterings," *Optics and Photonics News* 13 (3): 9 (March 2002); "Scatterings," *Optics and Photonics News* 16 (7): 9 (July/August 2003); Barry R. Masters, Review of *Secret Knowledge* by David Hockney, *Optics and Photonics News* 13 (6): 49 (June 2002).

6. There are references to something called a "Light Room," which was later Latinized as *camera lucida* in the work of Robert Hooke, but this is something altogether different. Claims that Johannes Kepler invented a *camera lucida* seem to be baseless.

{ 13 }

A Sea Change

Was Captain Nemo wary of what we might try in these frequented waters, or did he simply want to stay out of sight of the many ships of every nation plowing the Mediterranean? I don't know, but most of the time we cruised beneath the surface and far away from any shore. Either the Nautilus would come up just enough to let the pilothouse emerge, or it would dive to great depths.

—JULES VERNE, *20,000 LEAGUES UNDER THE SEA*[1]

The *Nautilus* would come up far enough to let the pilothouse (within what landlubbers like myself would call the *conning tower*, or American sailors call the *sail* or European ones the *fin*) emerge from the water because it lacked something that we would take for granted in a submarine—it had no periscope. Verne is often held up as the great extrapolator and predictor. And his submarine, the *Nautilus*, does have a great many accurate and clever devices—it has diving planes that help it maneuver underwater, a double hull and ballast tanks, and it's driven by electric motors (rather than steam or gasoline motors) that don't use up the precious store of compressed air. But his submarine had no torpedoes (it sank ships by ramming them) and no periscope. Why not? Surely the periscope is an obvious and simple bit of technology. How could Verne have overlooked it?

The answer, says translator and annotator Walter James Miller, is that the periscope isn't really so simple a technology. Verne, like the early submarine inventors, knew of its problems, and didn't trust it. Military submarines continued to use non-periscopic viewing ports into the twentieth century.[2] They started out that way, too. David Bushnell's *Turtle*, the first submarine used in a war (the American Revolution), was navigated by using a compass illuminated by the glow from luminous fungi in a piece of "foxwood." But on his return from his mission pilot Ezra Lee had to keep surfacing to get his bearings, and the odd motion of the unidentifiable craft attracted the attention of Hessians, who set out to investigate. Clearly, there was an advantage to using something smaller and less obtrusive as a viewing mechanism.

The simple periscope, consisting of two mirrors placed at 45° on a tube so that one can look over a wall, or similar obstruction, probably has a long history that has not been documented. It's hard not to believe that it has been invented and reinvented many times. Johannes Gutenberg was said to have sold such devices to pilgrims wanting to see over the heads of crowds before them back in the fifteenth century.

Certainly it was reinvented many times in the nineteenth century, when such devices were proposed for trench warfare (and were called *altiscopes*) and on surface Naval Vessels—Thomas Doughty used one on his ironclad the *Osage* in 1864. It was even proposed for a submarine as early as 1854, by French designer E. M. Marié-Davy.[3] During the American Civil War, one Professor Horstford of Cambridge Massachusetts designed a submarine with a periscope and an airlock.[4] In 1877 the Polish engineer Stefan Drzewiecki used one in his submarines, as did Claude Gorbet in France and Thorsten Nerdenfeldt in Sweden. The French submarine *Morse* had a periscope in 1899.[5]

Yet submarines continued to pass periscopes by in favor of viewing ports. There were several reasons for this. One of the most important is the fact that a periscope that is too long becomes useless without optics to magnify and to relay the image. In the first decade of the twentieth century, with little antireflection coating technology available, it wasn't practical to have a periscope system with any considerable length. (The value of a thin coating of magnesium fluoride was known, but coating technology was in its infancy.) The Fresnel losses at each lens surface would rapidly diminish the transmission to an unusable level. A second consideration was that, even at modest lengths, the periscope tube was subject to bending and, if the length matched the resonant frequency, the oscillations could make it unusable. This problem was much reduced with the development of an outer casing to stiffen the tube against water resistance, but some of the vibration came from the submarine itself. Third, the periscope had to be protected from moisture not only the omnipresent sea water (a problem made worse by the necessity of raising and lowering the periscope tube without introducing leaks) but also from the characteristically humid air *inside* the submarine. Processes and devices were needed for drying the interior of the periscope to keep the optics usable and free from condensation.

One center for serious development of periscopes was Britain. Captain Reginald Bacon, the head of the new British Submarine Service, thought that periscopes were essential for an operating submarine, and brought in the Irish optical scientist Sir Howard Grubb to design a periscope.[6] Frank Cable, of Electric Boat in Connecticut, noted the importance of the periscope and made it a priority in the United States, where Simon Lake took the lead in submarine design, including periscopes. Frederick Kollmorgen, who had designed telescopes in England and Austria, came to the United States and patented his first periscope in 1911. With the coming of First World War, and the stopping of commerce in optics with Europe, he set up the Kollmorgen Optical Corporation in Brooklyn, New York, specializing in periscopes. Prior to 1927, the only other major suppliers were Bausch and Lomb, Keuffel and Esser, and Kelvin, Bottomly, and Baird.

Periscope lengths were only 16–24 feet during First World War. Longer periscopes vibrated too much and lost too much light at each interface. By the time of the Second World War stronger structures and better coatings were available, lengthening the scopes to 30 feet. By 1960 they were up to 40 feet long.

Perhaps the biggest single contribution to developing periscope technology, as well as to lens design in general, was the development of thin-layer optical coatings. Chemical coatings had been employed since the end of the nineteenth century to reduce reflections, but there was too little control over thickness and few options for the coatings. Quarter wavelength Magnesium Fluoride coatings could reduce reflections to as little as 1 percent at the optimized wavelength. Grubb's or Kollmorgen's first periscopes had about 16 air/glass interfaces, and so optimum coating of this sort would transmit about 85 percent of the incoming light. Periscopes like this could be useful, but low light levels would compromise their effectiveness.

The technology that made thin films possible was the development in 1915 and the subsequent development of the diffusion pump. With this, high vacuum could be reliably and reproducibly achieved, and the long mean free paths necessary for vacuum deposition of thin films were possible. These were developed at Zeiss in Germany and General Electric in the United States. With antireflection coatings to minimize reflection losses at interfaces, designers were free to create multi-element lenses for cameras and other applications, and to craft relay telescopes for periscope tubes. Optical Glass formulation had undergone a similar burst in development in the last two decades of the nineteenth century, and now clear, bubble-free optical glass was available for thick lenses and for prisms to replace mirrors to reorient light.

Today, the periscope is disappearing. Since 1999, US Submarines have been designed with "Photonics Masts" instead. There is no need for a long and cumbersome periscope tube, which breaches the integrity of the submarine walls. Remote camera systems are easily placed on the top of long masts that do not require any particular spatial relationship to the Control Room.

It should be clear that no single person can take credit for the periscope. The idea is much older than the submarine, and, even when the idea was proposed, independently by many inventors, there were so many practical problems to implementing it that it was not until the end of a long period of inventing and refining that periscopes could be used as a dependable instrument. Nevertheless, some individuals have been singled out by zealous supporters as "the inventor of the periscope," including Marié-Davy, Simon Lake, Thomas Doughty, and others. One of the least probable is Morgan Robertson, an author of sea stories. (He gained notoriety as the author of *Futility*, in which a ship called the *Titan* sank under conditions eerily similar to those of the Titanic). When he died in 1915, his obituary claimed that he was the possible inventor of the periscope. He had described the device in his 1905 book "The Submarine Destroyer." Supposedly the directors of the Holland Submarine Company invited him in to talk. He said that he had patented a periscope model, and sold it to them for $50,000.[7] Another story, though, has it that Robertson was denied a patent because Jules Verne had already described the device in "20,000 Leagues Under the Sea."[8] None of these claims stand up—Robertson held no patent, and as far as we can tell never applied for one, and Verne did not describe a periscope, as we saw at the beginning of this piece.

Notes

Many thanks to John Merrill, who sent me a copy of his book *Looking Around: A Short History of Submarine Periscopes.* Please don't blame him for my conclusions or mistakes. I'd also like to thank Carl Floyd.

1. Jules Verne, *20,000 Leagues under the Sea*, trans. Walter James Miller and Frederick Paul Walter (1993), 234.

2. See Miller and Walter edition, previous reference, and also *The Annotated Jules Verne: 20,000 Leagues under the Sea*, with annotations by Walter James Miller (1976), 218–219.

3. Ibid., 218. See also the note 5 below.

4. Mark K. Ragan, *Submarine Warfare in the Civil War* (2001), 118–119. Ragan says that "Horstford appears to have been the first to propose [the use of a periscope] aboard a submarine," apparently unaware of Marié-Davy. Is this Eben Norton Horstford, Rumford Professor and Lecturer on the Application of Science to the Useful Arts at Harvard? He is better known for his reformulation of baking powder, and for championing the view that Vikings settled New England. His efforts are responsible for the statue of Leif Erikson that still stands on Commonwealth Avenue in Boston, and for the naming of Rumford Baking Powder, which is still being sold. See http://en.wikipedia.org/wiki/Eben_Norton_Horsford, accessed October 29, 2012.

5. John Merrill, *Looking Around: A Short History of Submarine Periscopes* (2002), 11–13.

6. Ibid., 15–17.

7. Wikipedia "talk" on Morgan Roberts, http://en.wikipedia.org/wiki/Talk:Morgan_Robertson, accessed September 22, 2008.

8. Jack Williamson, editorial, "Vanguard of Science," in *Startling Stories*, September 1939, reported on http://causticly-speaking.blogspot.com/2007/01/it-was-startling-that-fall.html, accessed September 22, 2008.

{14}

Thomas J. Pearsall and the Ultraviolet

The discovery of ultraviolet light is attributed to J. W. Ritter in 1801. He was following in the footsteps of William Herschel, who had noted that a thermometer placed in the region of a solar spectrum beyond the red caused the mercury to rise, thereby proving the existence of *infra-red* rays. Ritter looked for responses at the opposite end of the scale, beyond the violet. Nothing had resulted from placing a thermometer here, but Ritter placed paper that had been soaked in silver chloride there. It had been known for almost two centuries that silver chloride turned black on exposure to light, so he clearly hoped to be able to detect any unseen light with this aid. He did. Three years later Thomas Young, that indefatigable searcher for evidence of the wave nature of light, proved that the unseen rays would interfere with each other, recording the interference pattern on silver chloride paper, and thereby proving that ultraviolet was, indeed, a wave phenomenon.

In 1814 Joseph Fraunhofer started his work on the solar spectrum, noting the characteristic dark lines caused by absorption against the blackbody background, naming these, and measuring their wavelengths. They extended down to just below 300 nm, beyond which his glass prisms absorbed the ultraviolet radiation (although that was not to be known for many years). In 1842 George Gabriel Stokes replaced the glass prism with a quartz one, and showed that the rays extended down to wavelengths as short as 183 nm. At about the same time, John William Draper took a daguerreotype of the solar spectrum, and Leopoldo Nobili and Macedonio Melloni developed the thermopile to record unseen wavelengths. Alexandre-Edmond Becquerel also photographed the UV. So studies of the wavelengths of the light had begun. But the only good source of UV light was the sun, and nobody was using that for any investigative work. Mercury discharge lamps didn't become available until the end of the nineteenth century, and powerful UV sources did not come until the early twentieth century. So studies on the effects of ultraviolet light were comparative latecomers.[1, 2]

Or were they? While researching my doctoral thesis, I stumbled across a series of experiments that involved ultraviolet irradiation of materials conducted twelve years before that magic year of 1842. This treatment caused changes in ionization states and induced defect formation in crystals, which suggests pretty high levels of ultraviolet radiation. How was this possible at as early a date as 1830?

It was the work of Thomas J. Pearsall (1805–1883), a Laboratory Assistant at the Royal Institution[3] between 1827 and 1832. He assisted Michael Faraday and William Fox Talbot, working on chemical and electrical experiments.[4]

In 1830, Pearsall was conducting experiments on the effects of electricity on a form of Fluorite called *Chlorophane*, which had been found to glow green when heated, a process now called *Thermoluminescence*.[5] After heating, it lost this ability.

Pearsall found that passing a discharge over coarse fragments of chlorophane, or over powdered chlorophane, produced a brilliant green glow, as well. You could submit the same fragments to more discharges, and it would still glow. In fact, the glow seemed to get stronger with repeated discharges, whereas when you heated the crystal the glow eventually stopped, often accompanied by the crystal losing or changing its color. Whatever the heating did to the mineral, the discharge apparently undid it.

On the other hand, pieces of calcined chlorophane that were allowed to cool and "rest," or those exposed to sunlight in an effort to restore their thermoluminescence, did not exhibit that feature when heated again. There was something unique about the electrical discharge.

It's an odd experiment, and Pearsall never says what inspired it. It feels very much like the sort of experiment that might have been done 30 years earlier, when Alessandro Volta invented his *eudiometer*, a glass container in the form of a pistol with two wires coming in through the side and having a slight gap between them.[6] Volta used to fill the vessel with various gases, cork the top, and apply a Leyden jar to the leads to create a spark. Volta found that swamp gas produced a very satisfying "pop" as it burned, expanded, and blew out the cork. Here was Pearsall still experimenting with the same device three decades later, in an era when Faraday had moved beyond such simple spark experiments. I suspect that Pearsall just tried out the fluorite in his home-built eudiometer, observed the electrically induced fluorescence, and wondered if it might be related to the well-known phosphorescence that resulted from heating.

There certainly was some connection, since exposure to the electrical spark revived the glow, and the glow got brighter the more the crystal was exposed to the sparks. But the color of the glow from heating was green, while the glow caused by heating changed from blue-green to pink to white.

Different specimens, obtained from different areas, glowed with different colors when heated or treated with sparks. But again, the colors generally weren't the same. Pearsall dutifully recorded them all, describing the colors. Had spectroscopy been invented (beyond Thomas Melvill's early efforts—see chapter 10), he would undoubtedly have recorded the wavelengths.

Pearsall found that colorless fluorite, which did not glow at all when heated, could be given the ability to glow if exposed to several sparks from the spark container. "In this case, the property was conferred upon a substance which probably never possessed it previously," Pearsall wrote.

He found that exposing the fluorite to sunlight did *not* cause the fluorite to glow when subsequently heated, even when exposed as long as eight months.

So what was happening? Pearsall lacked many pieces of the puzzle, and much of the equipment that might have helped. One thing he was undoubtedly doing was exposing the crystals to bursts of ultraviolet light. Passage of electricity through the air breaks down the gases, which rapidly recombine with emission of photons. The main constituent of air is nitrogen, and the characteristic bluish color of sparks, be they static electricity sparks made by scuffing your feet on the rug and touching a doorknob, or by using a Tesla coil, or by a bolt of lightning, is mainly due to the recombination of ionized nitrogen. This produces a series of peaks centered around 360 nm, spaced about 20 nm apart. These long wavelength UV peaks are sufficiently energetic to induce many effects.

Pearsall later tried other substances—crystals such as apatite, lime carbonate, cuttlefish "bones," mother of pearl, scallop shells, oyster shells, marble. You get the feeling he was simply throwing whatever he had at hand into the *eudiometer* and seeing what happened.

What was one to make of all this? Pearsall observed that

> The colours of bodies, generally, are believed to be due to peculiar structures capable of decomposing light, and reflecting particular coloured rays.
>
> Since by experiment I have shown that colorific structure obtains in certain varieties of colourless fluors, as the result of intense electrization; and as electricity, under various conditions, may it not be advanced, that when matter (such as calcined fluor) which is not phosphorescent, is exposed to electrical discharges, that they cause vibrations of the particles, which, being repeated with each discharge, gradually modify the structure, and bring it into a peculiar state? May not the action of heat allow this state to return to what it was originally; and from the vibrations of the atoms of matter in changes of structure proceed the undulations fitted to produce light?

But he kept coming back to fluorite, noting the changes in different-colored fluorite from all over the world. Eventually, he found something very interesting. He could produce color in previously uncolored fluorite by exposing it to enough sparks. "There appears every reason to conclude that the colour induced is the effect of structure alone," he wrote (and not of impurities added to the crystal).

Pearsall was correct—ultraviolet light (and higher energy photons) are well known for their ability to produce defects, and in particular the defects called *color centers*, which can have absorption bands in the visible. There are many types of defects in crystals, but one of the more interesting—and relevant, in this case—is the *color center*. These are defects with a simple structure, that can cause a normally colorless crystal to selectively absorb light, and thus to become colored, hence, the name. The simplest sort is what is called the *F-center* in simple salt crystals.[7] In sodium chloride, the sodium and chlorine atoms are arranged in planes, each of which has the structure of a checkerboard, with, say, chlorine atoms as the black squares and sodium atoms as the red squares. Subsequent layers have the same structure, but with reds and blacks reversed. A chlorine atom is surrounded on four sides by equidistant nearest-neighbor sodium atoms, with one on each side (East, West, North, South), as well as one in the plane above and the plane below. Each sodium atom is itself surrounded by six chlorine atoms in just the same way. These sodiums and chlorines are not actually atoms, but ions—each sodium has lost an electron and each chlorine has gained one, making them positively and negatively charged, respectively. The electrostatic force between these oppositely charged ions holds the crystal together.

No crystal is perfect, and there are defects throughout, but most of the ions really do form a nearly perfect checkerboard pattern of alternating sodiums and chlorines. Where there is a missing chlorine, the space left has a net positive charge, and an electron may be trapped there, rattling around (as some early researchers liked to put it) like a pea in a muffin tin. This is the basic F-Center. Until the advent of modern nanomachining techniques, and the development of Quantum Dots, F-Centers were probably the closest thing to the Quantum Mechanical ideal of a Square Well. A color center has similar discrete energy levels whose separations vary nearly as the inverse square of the size of the crystal. There are many other types of color-absorbing defects—pairs of F-centers, and sets of three, or F-centers next to impurity ions in the crystal, and so on.

It turns out that an easy way to create F-centers is to expose the crystal to ionizing radiation. Gamma rays and X-rays reliably create color centers, but so can ultraviolet rays, if they are short enough or intense enough. UV rays can also be used to turn one type of color center into another. I have done much of this work myself in various types of alkali halide crystals.

But such color centers don't only appear in alkali halides (sodium chloride, lithium fluoride, potassium bromide, etc.) They also appear in alkaline earth chalcogenides (one step in each way on the periodic table—Calcium Oxide and others), as well as in other crystals, including fluorite and diamond. As far as I can tell, Thomas Pearsall was the first person to create color centers in originally colorless crystals by exposing them to ultraviolet radiation. Ultraviolet light tends to be absorbed best near the surface, and the highest concentrations occur at edges and points, where two or more surfaces meet and the radiation is the most intense. So Pearsall's observation of heaviest coloring at edges makes sense in terms of what we know today. Furthermore, the centers can be persuaded to dissociate under heating, which gives the electrons enough energy to escape the potential energy well it is sitting in. Often simply pressure is enough to do the same thing, an effect called *triboluminescence*. This can be spectacular for some alkali halide crystals, but a more familiar manifestation of the effect is the way that you can see wintergreen Life Savers candy "glowing" if you chew them in a dark room in front of a mirror. Pearsall, sadly, did not know about this.

Crystals with colors due to defect centers have long been known—smoky quartz is one, and the blue potassium chloride of Northern Africa described by

Herodotus is another, but as far as I am aware Thomas Pearsall was the first to create such color centers by deliberate irradiation of colorless crystals by exposure to what was (although he did not know it) Ultraviolet Radiation.

After publishing these papers, Pearsall does not appear to have done much independent work. He seems to have contributed to the 1839 paper *Notices of the Effects of Lead upon the System* and wrote *On the Red Solutions of Manganese.*[4] He served as Laboratory Assistant in the Royal Institution from 1827 to 1832 and as Chemical Assistant there at some point. He worked with William Fox Talbot, the optical physicist and one of the inventors of photography, and may be the same Thomas Pearsall who served as Secretary of Birkbeck College at the University of London from 1863 to 1866 or thereabouts. His name appears as a secretary for the Chemical Society occasionally in the 1850s, and he apparently wrote on various topics from time to time. In *The British Journal of Photography* for November 9, 1877, he is reported to have delivered a lecture on *Educational Aids of Photographic Exhibits* to the South London Photographic Society.[8] But there don't seem to be any other physical or optical papers of any length. Mr. Pearsall was one of many contributors to the enterprise of science who made this one mark, at least, and has the distinction of an interesting "first."

Notes

1. Lewis R. Koller, *Ultraviolet Radiation,* 2d ed. ([1952] 1965), 1–5.

2. James A. R. Samson, *Techniques of Vacuum Ultraviolet Spectroscopy* (1967), 1–4.

3. Thomas Pearsall's two papers on electricity and thermoluminescence are "On the Effects of Electricity upon Minerals which are Phosphorescent by Heat," *Journal of the Royal Institution of Great Britain* 1: 77–83 (1831) and "Further Experiments on the Communication of Phosphorescence and Colour to Bodies by Electricity," *Journal of the Royal Institution of Great Britain 1*: 267–281 (1831).

4. Pearsall's work on the "Red Solutions of Manganese" appears in the *Journal of the Royal Society of Great Britain* 2: 49–64 (August 1831). The work on Lead is in James Alderson and Thomas J. Pearsall, "Notices of the Effects of Lead Upon the System," *Medico-Chirurgical Transactions* 22: 82–94 (1839). On him as Secretary of Birkbeck, see http://www.bbk.ac.uk/about-us/governance/former-officers, accessed June 2012. His association with Talbot is given in a letter from Talbot to Michael Faraday in March 1832 reproduced here http://foxtalbot.dmu.ac.uk/letters/ftbh.php?docnum=02313, accessed June 20, 2012.

5. http://www.mindat.org/min-948.html, accessed May 29, 2012.

6. James Burke, *Connections* (1978), 177–179.

7. Karl Przibram, *Irradiation Colours and Luminescence* (1956) seems to have been the first to apply the name *FarbeZentrum* ("Color Center") to these irradiation colors, from which the English term *F-Center* for one particular variety of such defect centers was in turn derived.

8. Thomas J. Pearsall, *The British Journal of Photography* 24: 537 (November 9, 1877), available online at http://books.google.com/books, accessed October 10, 2012.

{ 15 }

If at First You Don't Succeed . . .

James Burke, in his PBS series *Connections*, observed that the history of science and technology is all too often treated as if it were a simple story of unimpeded progress, each step inexorably and inevitably leading to the next, so that our knowledge constantly increases in a direct and flawless fashion. Textbooks reinforce this view, telling the history of science in a linear fashion, going from discovery to successful theory in an apparently effortless story of unfolding knowledge.

Real life doesn't work that way. There are misconceptions and mistakes. There are long periods when no progress in understanding on a topic at all is made. There are heartbreaking cases where a discovery is made, only to be lost again, or credit is misattributed. All of this is too confusing and complex to be told in introductory texts, however. It's one of the reasons to keep studying one's subject and rereading the histories. One of the delights of Burke's work is seeing how unexpected and haphazard progress is. Stephen Jay Gould, Martin Gardner, and others have written about the lack of good reporting of mistakes, failures, and dead ends. There should be more coverage of the errors and false starts of science. They teach us how to pursue the truth, how to recognize a wrong turning when we find it, and hearten us when we feel that we have reached such a turning ourselves.

Consider the early history of the first great triumph of spectroscopy—unraveling the spectrum of the hydrogen atom. The hydrogen emission spectrum, with its very regular series of lines in the visible spectrum, was first given accurate wavelength values by Anders Ångstrom and Julius Plücher of Uppsala in 1853. Of all the observed spectra, hydrogen, with its very basic structure, seemed to be the one most likely to have a simple mathematical progression to its lines. As the story is almost always told, everyone expected elemental spectra to consist of a fundamental frequency and its overtones, which would be integral or half-integral multiples of that fundamental. This is the way systems such as mechanical vibrations behave. For instance, a simply supported beam, supported without being clamped at each end, vibrates at the fundamental mode with mixtures of the overtones, which are integral multiples of the fundamental frequency. The simply supported beam is the basis of an entire class of musical instruments—the xylophone, glockenspiel, marimba, etc., which have pleasing musical tones. Alas, the observed wavelengths and frequencies of the hydrogen atom did not follow such a progression.

It remained (so the story continues) for a humble schoolteacher, J. J. Balmer, who had no such biases about the nature of the hydrogen atom's frequencies, to look at the data afresh and to discover a purely mathematical relationship that fit the data perfectly in 1885. Balmer did not himself understand the nature of the fit, but neither did anyone else, until Niels Bohr formulated his Quantum Theory.

The story is practically the only way this important bit of spectroscopic theory is told, but it glosses over those failed attempts that filled the years between 1853 and 1885, and it doesn't quite get the rest of the story correct, either.

I was interested in this aspect of history because I vaguely recalled hearing that, in the interval, someone had actually come up with a formula that fit the observed hydrogen spectrum, and did it perfectly. But it was erroneous, a dead end, and it represented a false lead in the history of science. What was this formula? How close was it? Was it perhaps an approximation to the correct formula, or was it, indeed, a completely incorrect guess that managed, against the odds, to perfectly fit the observations? If so, how did they know that it was false, and stop pursuing it?

After some searching, I found what I was after.[1] The characteristic spectral lines that provided a unique set of fingerprints for the elements (and provided a powerful tool for identifying new elements) also provided an exasperating and perplexing riddle. There seemed to be no obvious explanation or trend for the numbers, intensities, and especially the spacings of the lines. But the lines were not without some trends that were noticed. In 1861–1862 Alexander Mitscherlich of the University of Berlin noted that the spacings of the two principal lines in the spectra of barium chloride, barium iodide, and barium bromide were similar but varied as the atomic weight of the compound.[2] There was a relationship between the halides of the other alkaline earths, calcium and strontium, but this was an *inverse* relationship. There was clearly a connection of some sort, but it was frustratingly obtuse. François Lecoq de Boisbaudran, an unaffiliated scientist, noted in 1869 that the principal lines of potassium had spacings that increased in nearly geometrical progression[3]—11.9 to 16.1 to 23.2.

The most striking fitting, however, which must certainly be the one that inspired the legend I'd heard, was the work of George Johnstone Stoney, professor of Natural Philosophy at Queen's University in Dublin. Photographs show Stoney (1826–1911) with a patriarchal beard. If you look him up in the reference books little mention of this spectroscopic work is made. His real claim to fame is his pioneering work on the electron, for which he coined the name. Even here, however, he is overshadowed by Thompson.

Stoney, too, imagined that spectra must consist of harmonics of some fundamental frequency. Yet no obvious progression existed. Stoney's idea was that not all lines might be present, that some might be exceptionally weak or even absent, due to some sort of interference effect. He therefore tried to find numerical ratios between observed lines, and in the case of Hydrogen, found both exceptional agreement and perverse anomaly.

The four visible emission lines of hydrogen were measured at 4102.37 Å, 4341.68 Å, 4862.11 Å, and 6563.93 Å (Stoney corrected the measured values to obtain the wavelengths in vacuum). Three of these (excepting the second) he found to be in the ratio of 20: 27: 32, indicating a fundamental vibration[4] of 131,277 Å. He doesn't give the details of how he discovered this, but the agreement is phenomenal. Maddeningly, no other lines fit the series. Equally maddeningly, the line at 4341 Å does NOT fit the pattern.

Looking to bolster his case, Stoney collaborated with J. Emerson Reynolds, Keeper of the Mineralogical Department of the Royal Dublin Society to measure and fit the absorption spectra of chlorochromic anhydride, today called *chromyl chloride*. Exactly why this aggressive oxidizer was chosen is not stated, but it could be because the substance, a liquid at room temperature, does not react with itself except through van der Waals forces, as evidenced by its low melting point. Its spectrum would not be confused by self-interactions.

Stoney and Reynolds fit 31 lines of the substance, but not as well as in the case of hydrogen. It also required very high harmonics,[5] up to the 733rd. This iffy agreement of the chlorochromic anhydride and the high numbers required gave many pause, but J. L. Soret, editor of the *Archives des Sciences Physiques et Naturelles* was impressed with, as he put it, the "extreme precision," and tried his own hand at fitting the spectra of magnesium and cadmium.

Yet a complete and reliable theory explaining the spectral lines and the evident gaps did not emerge. Something was clearly wrong. The entire issue of such high-order overtones was addressed by Franz Arthur Friedrich Schuster, a German-born pioneering spectroscopist (who coined the word "spectroscopy") at the University of Manchester. In "On Harmonic Ratios in the Spectra of Gases," he explored the probability of such close fits between ratios of observed spectral lines occurring purely by chance[6] for ratios of numbers less than 100. Schuster was able to show that it was certainly possible to find ratios of pairs of numbers less than 100 that fit the observed spectra for sodium, copper, barium, and iron, and that the fits are no better than one would expect by chance. There was therefore no compelling need to see an underlying physical law behind such coincidences.[7]

Schuster's report had a chilling effect on the search for harmonics. At the June 12, 1885, meeting of the Berlin Physical Society, it was stated that the attempt to find simple harmonic relations between spectral lines was abandoned after Schuster's discussion.[8]

Johann Jacob Balmer (1825–1898) was, it is true, a teacher at a secondary school for girls in Switzerland. But he was also a lecturer in mathematics at the University of Basel, so he was not quite as obscure as he has been portrayed. Furthermore, in spite of the legend, he was not unaware of prior work. He explicitly mentions Stoney, his whole-number ratios corresponding to ratios of hydrogen wavelengths, and of his search for higher harmonics of a supposed fundamental frequency.

As others had been, Balmer was distrustful of the large numbers of the harmonics often needed to obtain fits. But the low number ratios in the case of hydrogen intrigued him and convinced him that some simple law must be hiding there.[9] Balmer noted that Stoney's ratios for the lines could be expressed as whole-number ratios of 9/5, 4/3, and 9/8.

He was able to express the fourth visible line as a ratio as well. Stoney had been looking for whole-number harmonics and had missed the possibility of ratios of whole numbers. The ratio was 25/21, which, although larger, was still a ratio of relatively small numbers (certainly smaller than the multiples of a fundamental that had to be measured in hundreds that Stoney had required). The fit of this fourth line to the ratio was as good as the others. Multiplying the numerator and denominator of two of these ratios by four gave the values:

$$\frac{9}{5}, \frac{16}{12}, \frac{25}{21}, \frac{36}{32}$$

The numerators form an ascending sequence of squares. Moreover, the denominators are all smaller than the numerators by four (which, Balmer observed, is also a square). Balmer concluded that he had found a formula that had a fundamental frequency—he called it a "key note"—of 3645.6 Å, but in which the harmonics were not simple multiples, but followed the relationship:

$$\lambda = 3645.6 \frac{n^2}{n^2 - 4} = 3645.6 \frac{n^2}{n^2 - 2^2}$$

This is the Balmer Formula, which can be rearranged to give the Rydberg-Ritz formula, and which was confirmed by Bohr's atomic model. Balmer saw the possibilities to extend the formula by inserting other values of **n**, and of substituting other squares for the 4 in the denominator. He used the formula on the ultraviolet hydrogen lines measured by astronomer William Huggins and H. W. Vogel in the 1870s and found very good agreement, to within 1 Å for Huggins's values and no more than 3.9 Å for Vogel's. The error was not with the formula, but the measurements. Improved precision in data newly measured by Alfred Cornu of Paris and Gustav Muller of Potsdam were fitted by Hagenbach in 1886 and yielded much better fits.[10]

Balmer's formula doesn't agree with the form of harmonics in vibration or acoustics. Balmer himself saw it as a sort of Pythagorean progression, and the denominator does indeed resemble the equation for one of the legs of a right triangle. There was no theory yet which explained this odd relationship, but the goodness of the fit and the predictive power of Balmer's formula had been proven.

What's interesting here is that Stoney did indeed stumble upon an important relationship, but his own model of high-order harmonics with missing orders was not correct, and led him to pursue other, less convincing fits. Schuster was quite right in dispensing with Stoney's chlorochromic anhydride and Soret's data with his statistical argument. In hindsight, it's easy to say that he went too far in dismissing Stoney's work on hydrogen as well, which agreed far better with the simple ratios than the other work, and which used harmonic numbers much smaller than the other fits that had been attempted. It was necessary to jettison the approximate fits, which were the result of coincidence, but the hydrogen spectrum data "baby" was thrown out with the bathwater. It remained for the unaffiliated Balmer, who had not been dissuaded by these arguments against harmonics, to take a fresh look (but an informed one, familiar with prior work, and not the naïve look of legend) and to discover the law of progression.

Notes

1. This material is drawn from Klaus Hentschel, *Mapping the Spectrum* (2002); William McGucken, *Nineteenth Century Spectroscopy* (1969); Chris Candler, *Atomic Spectra and the Vector Model* (1964); and from other sources listed below. There is also a brief account, with good quotes from Schuster and Balmer, in the unattributed paper "The Balmer Formula," which has been at several websites, including http://www.chemteam.info/Electrons/Balmer-Formula.html, accessed November 5, 2012.

2. McGucken, *Nineteenth Century Spectroscopy*, 103–104.

3. Ibid., 106–107.

4. J. G. Stoney, "On the Cause of the Interrupted Spectra of Gases," *Philosophical Magazine* 41: 291–296 (1871).

5. J. G. Stoney and J. E. Reynolds, "An Inquiry into the Interrupted Spectra of Gases, Part 2. On the Absorption Spectrum of Chlorochromic Anhydride," *Philosophical Magazine* 42: 41–52 (1872).

6. First, a brief notice in *Nature* 20: 533 (1879), then expanded into a longer article "On Harmonic Ratios in the Spectra of Gases," *Proceedings of the Royal Society of London* 31A: 337–347 (1881).

7. McGucken, *Nineteenth Century Spectroscopy*, 122–125.

8. *Nature* 32: 312 (1885), cited in McGucken, *Nineteenth Century Spectroscopy*, 125.

9. Balmer's work appears as "Notiz über die Spectrallinien der Wasserstoffs" in the obscure *Verhandlungen der Naturforschenden Gesellschaft in Basel* 7: 548–560, 750–752 (1883), but more accessible is the shorter article in *Annallen der Physik* 25: 80–87 (1885). Also of importance is the note by A. Hagenbach in *Die Naturwissenschaften* 9: 451–453 (1921) by the son of Balmer's collaborator, Professor Jacob Eduard Hagenbach-Bischoff.

10. Hentschel, *Mapping the Spectrum*, 298–299.

{16}

More Than a Burner

Several years ago I ran across a cartoon by Sidney Harris, the scientifically oriented artist whose work appears in venues such as *American Scientist* and *Scientific American*. It shows an obviously nineteenth-century scene of a man in a top hat addressing another man at the table. "Bunsen," says the top-hatted man, "I must tell you how excellent your study of chemical spectroscopy is, as is your pioneer work in photochemistry—but what really impresses me is that cute little burner you've come up with."

I was amazed. I have to admit that all I really knew about Bunsen was that burner—I didn't even know his full name. I'd used that natural gas burner in college and high school, and (with uncles and cousins in chemistry) I knew of it well before that. But I never gave a thought to Bunsen's real work. This cartoon was the first I'd heard of Bunsen as a spectroscopist or electrochemist.

So what is the story on Bunsen, and what does he really have to do with that burner?

Robert Wilhelm Bunsen was born in 1811 in Göttingen, Germany, the youngest of the four sons of Christian Bunsen, professor of philology at the University there. Robert studied chemistry at Holzminden and Göttingen, then toured European laboratories for three years before returning to his home town to be a lecturer at the University before going on to positions at Kassel, Marburg, and finally Heidelberg.[1]

He made a false start in organic chemistry, doing work on the arsenical compound cacodyl, but soon settled in inorganic chemistry. He lost an eye due to an explosion of cacodyl cyanide. For about ten years, he was heavily involved in the study of gases. Then Bunsen's studies evolved around galvanic batteries, and he invented the Bunsen battery, which used inexpensive carbon in place of platinum or copper as the negative pole. He used this in electrochemistry, producing sodium, aluminum, and other metals from their chlorides.

His work in spectroscopy, our reason for covering his work, did not begin until comparatively late, in the 1860s. With his younger protégé Gustav Kirchhoff, they observed the colors and spectra of chemical salts, heated using an alcohol flame or burner flame.[2] They also observed that placing such a flame in a white light lead to dark absorption bands. (In addition, they produced spectra by passing electrical current through a Ruhmkorff coil discharge, with a small sample of the material under test applied to the electrodes.) Both the light and dark bands were characteristic of the metallic part of the salt, they found, and the test was sensitive enough to require only a tiny amount of the test material (although it had to be highly purified to produce only the characteristic lines). They were able to identify cesium and rubidium from extremely small samples (in 1860 and 1861), and subsequently isolated larger quantities of the elements by more conventional means.

Bunsen was a dedicated teacher, lecturing for 100 hours each year through seventy-four semesters. He designed and built his own apparatus, and was particularly skilled at glassblowing. He was reputed to have tough skin on his hands, and he used to demonstrate this by holding his hand in the flame of his burner, and he used to remove the lids of hot crucibles with his bare hands.

Bunsen never married. The story (probably not true) was told that, while a young man at Marburg, he had proposed to a young woman and had his suit accepted, but he got so absorbed in his study of cacodyl that he neglected to visit her for several weeks, and became uncertain whether or not he had actually proposed. He visited her without apologizing for his absence and re-proposed, with predictable results. Years later, Kirchhoff's wife asked Bunsen why he had never married. "Heaven forbid," he replied, "When I return at night, I should find an unwashed child on each step."

What about that famous burner?[3] The one he used for his spectroscopic work, and which he used to impress his students with, by holding his hand in it? Some people claim that he didn't invent it—his technician did, or he simply adapted it from Faraday, or in some other way he is credited with one thing that, ironically, he really didn't have much to do with. But Bunsen was reported to be a superb experimentalist, who invented the Bunsen battery, a much improved eudiometer for gas measurements, the grease-spot photometer for light measurements, the highly sensitive spectrometer that he and Kirchhoff used in their measurements, an ice calorimeter, a vapor calorimeter, a thermopile, and a vacuum pump. He stressed the importance of building one's own experimental apparatus to his students and complained when they did not achieve proficiency at glassblowing.

All of which supports the notion that he improved the burner to achieve the one that now bears his name. Gas burners using the newly available coal gas had been built by Michael Faraday in England and by Aimé Argand in Switzerland, which had a system of delivering the gas for lighting before Germany did. Bunsen did not invent the gas burner, but I doubt if anyone ever claimed that he did. The gas system that came to operate in Germany apparently did not have the same flow rate as in England. Henry Roscoe, one of Bunsen's English students, brought a lamp called a "gauze burner" to Heidelberg, but at low flows it tended to go out. Bunsen experimented with different tube diameters, aided by his laboratory assistant, Peter Desaga, finally coming up with something very much like the present-day “Bunsen Burners,” although it lacked the rotating ring that adjusted the flow of air to the flame.

But the burner Bunsen and Desaga modified did what it needed to do—it provided a colorless, sootless flame that was ideal for spectroscopic work and general laboratory applications. Bunsen himself did not patent the device. Desaga built many copies for the laboratory, then, with Bunsen’s approval, he built more for sale to other laboratories, calling them “Bunsen’s Burners,” and defending them against others who tried to file patents on the device.

Desaga’s family continued in the scientific instrument business and was still manufacturing the burners into the 1950s. The firm is still in business, but a perusal of its website fails to turn up any of the burners, many of which seem to be manufactured in Asia today. In the modern laboratory the burner uses not coal gas (made by heating coal or coke in air, producing a mixture which was about 50 percent hydrogen, with additions of carbon dioxide, carbon monoxide, methane, and nitrogen, the proportions depending upon the process used), but natural gas (methane, mostly, but with other gases present, including butane, propane, carbon dioxide, nitrogen, hydrogen sulfide, and even helium).That it was able to make the change is a fortuitous tribute to the man’s inventiveness, guaranteeing that his name would live on, even among people unaware of his many other and arguably more significant accomplishments.

Notes

1. On Robert Bunsen, see *Dictionary of Scientific Biography* (1970), Vol. 2, pp. 586–590; *Physics Education* 34 (5): 321–326 (1999); the Wikipedia entry at http://en.wikipedia.org/wiki/Robert_Bunsen, accessed October 7, 2012; and most especially *Journal of Chemical Education* 4 (4): 431–439 (1927), for which the author interviewed some of Bunsen’s former students.

2. On the place of Bunsen and Kirchhoff in the history of spectroscopy, see William McGucken, *Nineteenth Century Spectroscopy* (1969), especially pp. 26–31.

3. The history of the Bunsen Burner gets treated at irregular intervals in *The Journal of Chemical Education*. See *Journal of Chemical Education* 9: 1963–1969 (1932); 27 (9): 514–515 (1950); 33 (1): 20–22 (1956); 77 (5): 558–559 (2000); 82 (4): 518 (2005). See also http://en.wikipedia.org/wiki/Bunsen_burner, accessed October 7, 2012.

{17}

Apply Light Pressure

I first encountered the Crookes radiometer as a child. I saw a display consisting of a couple of dozen radiometers, light bulb-shaped things with odd little anemometer-like vanes inside, sitting in tiers atop shelves under bright lights and rotating madly. I had no idea what it was supposed to be, or how it worked, but it was very impressive.

Years later, in high school, my physics teacher explained the principle of the operation of the radiometer. It worked like an anemometer, he said. The spinning vanes were sealed into a bulb from which all the air had been evacuated, so that nothing could hinder the rotation of the vanes. Photons impinged on the square panels that were the vanes, but one side of the vane was colored white, or had reflecting material on it, and the photon rebounded, transferring twice its initial momentum to the vane. But photons hitting on the black side were absorbed and only transferred their momentum a single time. The vanes thus rotated preferentially away from the white side, with the black side being pushed toward the brightest source of light.

But when you looked at the radiometer, you saw that it rotated in precisely the *opposite* fashion, with the black sides receding from the source of light, and the white sides approaching the light. Something was clearly wrong with the explanation. In fact, said my teacher, he had lied. The radiometer really rotated in what looked like the Wrong Direction. The *real* explanation was that the motive power for the vanes didn't come from photons at all. The vacuum in the bulb was not perfect, and there was a little residual air. When the light fell on the vanes, the white or mirrored side didn't absorb any, but the black side absorbed it all, and got hotter. When air molecules struck the surface, they picked up extra energy and momentum from the hotter black side than they did from the cooler white side, and so they rebounded more from the black side, which in its turn (because of Newton's Third Law—for every action there's an opposite and equal reaction) rebounded with more momentum than the white side did. And therefore, the vanes of the radiometer rotated with the black sides going away from the light source, as observed.

What I didn't know at the time was that this explanation has been known to be incorrect for over a century. This is rarely pointed out in classes and lectures, so most people don't realize this, even though there have been several excellent articles on the topic.[1]

There had been attempts to measure the momentum of light going back well before William Crookes came up with his radiometer. Jean-Jacques d'Ortous De Mairan, physicist, astronomer, and later Permanent Secretary of the Académie Royale des Sciences, assisted by chemist (and superintendent of the Royal Gardens) Charles François de Cisternay du Fay[2] tried to observe it in 1747, and in 1796 the Reverend Abraham Bennett constructed a device similar to the one Crookes would eventually build, but without as good a vacuum.[3] Augustin Fresnel seems to have had some success with a device in 1825, but he died shortly after publishing, and his work was little noticed.[4] In some cases these were intended to demonstrate the corpuscular theory of light. They were not successful, mainly because the pressure that was to be measured was itself so small. Even James Clerk Maxwell, who propounded the electromagnetic theory of light, but knew that it implied that it would be capable of generating pressure, built an apparatus to try to observe light pressure, but without success.

William Crookes was a chemist who was well enough off to fund his own laboratory. In 1873 he attempted to measure the weight of a tiny sample of the element thallium (which he had himself discovered 12 years earlier) and found that it was difficult to obtain a consistent reading. He used the newly invented Geissler mercury vacuum pump to evacuate the air from his chamber, but found that the weight seemed to depend upon the temperature of the sample, and that it made a difference from which direction he heated it. Experimentation showed that blackened surfaces were repelled more than silvered ones. Crookes demonstrated the effect of repulsion before the Royal Society of London in 1874. His published work appeared in the *Journal of the Royal Society of London*, and it ran for a surprising 26 pages.[5] Crookes would ultimately publish five articles on the effect and its variations over the next five years. In 1876 he devised the simple and elegant device with four vanes blackened on one side but silvered on the other. He named it the "Radiometer" or "Light Mill," both terms still in use today.

At first, Crookes believed that the observed sense of rotation was correct—that black really should recede from the source of light, since the black side absorbed the photon and its momentum, but the white side clearly did not. Fluid mechanics pioneer Osborne Reynolds (who had been present at the first demonstration) saw that the effect was consistent with heating, rather than incident light. His student, Arthur Schuster, performed an experiment in which a radiometer was suspended from fine wires and illuminated. The radiometer vanes and the radiometer casing rotated in opposite directions, following Newton's laws of action and reaction, indicating that something within the case was being impelled in the opposite direction to that of the vanes. This, Reynolds explained, must be residual gas producing the effect, and produced the kinetic gas explanation that my physics teacher later passed on to me.

Crookes was forced to admit that he was not, in fact, measuring radiation pressure of the sort that Maxwell had just proposed. But Crookes still didn't properly understand the mechanism responsible. He thought that the mean free path of the gas molecule ought to be larger than the size of the bulb, and that the molecule thus managed to "push off" one wall of the bulb and transfer the momentum thus gained to the vanes of the radiometer. Crookes doesn't seem to have properly come to grips with the true explanation of the radiometer, but this work on long mean free paths of particles in an evacuated bulb striking a target within that bulb lead him directly to the creation of the equally famous Crookes Tube, which made cathode rays visible, except where blocked by his Maltese Cross target.

Both Maxwell and Reynolds wrote papers regarding the true nature of the impulsive force, but these were heavily mathematical.[6] A more accessible explanation was offered by Sutherland in 1896, but it was largely overlooked. Interest in the subject revived in the early 1920s, with a series of papers by various physicists (including Albert Einstein). The problem with the simple kinetic gas explanation is that those gas molecules are not so sparse that they can be assumed to be non-interacting. In fact, gas molecules rebounding from the surface of the vane interfere with incoming molecules, and the hotter the surface, the more effective the rebounding molecules are in blocking incoming molecules. The net effect is that the forces equal out, and the vanes ought not to rotate. Dewar and Tait pointed this out, although in an obscure way, in 1875. Reynolds came up with a mental picture that made the argument more intuitive. Imagine two parallel plates attached together with low-pressure gas filling the space between them. If one plate is warmer than the other, the construction should accelerate in the direction of the warmer plate, with no external forces acting.

The reason that the vanes do turn, despite this apparently arresting effect, is that the argument breaks down near the edges of the finite vanes. The outgoing molecules no longer deflect the incoming ones as much on the hotter side than the cooler one, and the imbalance in forces allows the vanes to rotate.[7]

The Crookes radiometer works for pressures in the range 10 to 100 microns pressure. Blum and Roller report that true radiation pressure was finally observed in a standard Crookes radiometer that had been evacuated to very low pressure, but the effect of radiation pressure had been observed in different apparatus by Pyotr Lebedev[8] of Moscow University in 1899 and by E. F. Nichols and G. F. Hull[9] of Dartmouth in 1903.

The angular momentum of light was measured in 1936.[10] More recently, a nanoscale light mill was constructed and driven with polarized laser light by Xiang Shang et al. at the Lawrence Berkeley lab and the University of California at Berkeley.[11]

In an interesting variation, the effect of radiation pressure using a Crookes-type device was used to demonstrate *acoustic* radiation pressure in an experiment by

Bruce Denardo and T. G. Simmons[12] in 2003. Operating at room pressure, the device rotated in the "correct" direction. Even though the edge effect was present, the mean free path is very small, and so the effect is negligible. For the first time, an accessible and relatively simple device demonstrated the effect of radiation pressure directly.

The impression one might get from this brief account is that Crookes was not a very competent scientist, but, as A. E. Woodruff notes,[13] "Crookes was no mathematical physicist, but a first rate experimentalist." Recall that his was the first such device to produce a positive result when others, including Maxwell, had failed. It is surely a fitting tribute that his radiometer still enjoys brisk sales at scientific stores and museum gift shops, long after most of its contemporary devices have long been forgotten.

Besides his discovery of Thallium, Crookes made a number of other discoveries and became a Fellow of the Royal Society in 1863, and was knighted in 1897. Characteristically, he put a depiction of his radiometer on his arms and chose as his motto *Ubi Crux Ibi Lux*—"Where the Cross is, There is Light." Of course, with just a little imagination, it could be read as "Where Crookes is, There is Light," and that clever dual-language pun alone makes him worth quoting.

Notes

1. The story of the Radiometer has been told several times in recent years. Undoubtedly the best and fullest account is by Arthur E. Woodruff. He gives it in "The Radiometer and How it Does Not Work," *The Physics Teacher* 6 (4): 358–363 (1968) and in "William Crookes and the Radiometer," *Isis* 57 (2): 188–198 (1966), and I have drawn heavily on these. See also "Trick of the Light," *New Scientist* 183 (2461): 48–49 (August 21, 2004); Clifton W. Draper, "The Crookes Radiometer Revisited," *Journal of Chemical Education* 53 (6): 356–357 (1976); and John Witzel, "The Radiometer: A 130-Year-Old Mystery," *IEEE Instrumentation and Measurement Magazine* 5 (3): 60–61 (September 2002). See also S. G. Brush and C. W. F. Everett, "Maxwell, Osborne Reynolds, and the Radiometer," *Historical Studies in the Physical Sciences* **1**: 105–125 (1969). On the early attempts to measure light pressure, see Morton L. Schagrin, "Early Observations and Calculations on Light Pressure," *Am. J. Phys* 42 (11): 927–940 (November 1974); and John Worral, "The Pressure of Light: The Strange Case of the Vacillating 'Crucial Experiment,'" *Stud. Hist Phil. Sci.* **13** (2): 133–171 (1982), online at http://www.colin-baxter.com/academic/bib/downloads/worrall82.pdf, accessed October 3, 2012.

2. C. F. duFay, "Eclaircissemens sur le traité physique et historique de l'aurore boréale," *Histoire de l'Académie Royale des Sciences* 363–435 (1747).

3. A. Bennett, "Experiments on a New Suspension of the Magnetic Needle," *Philosophical Transactions of the Royal Society* 87 (1796).

4. A. Fresnel, "Note sur la repulsion que des corps échauffés execent sur les autres à des distances sensible," *Ann. De Chímíe et de Phys.* 29 (2): 57, 107 (1825).

5. William Crookes, "On Attraction and Repulsion Resulting from Radiation," *Phil. Trans. of the Royal Soc. of London* 164: 501–527 (1874).

6. J. Clerk Maxwell, "On Stresses in Rarefied Gases Arising from Inequalities of Temperature," *Proc. Royal Soc. of London* 27: 303–308 (1878), online at http://rspl.

royalsocietypublishing.org/content/27/185-189/304.full.pdf, accessed October 3, 2012; Osborne Reynolds, "On the Force Caused by the Communication of Heat between a Surface and a Gas, and on a New Photometer," *Proc. Royal Soc. of London* 24: 388–391 (1876), online at http://rspl.royalsocietypublishing.org/content/24/164-170/387.full.pdf, accessed October 3, 2012, and, immediately following, Arthur Schuster, "On the Nature of the Force Producing the Motion of a Body Exposed to Rays of Heat and Light," *Proc. Royal Soc. of London* 24: 391ff. (1876).

7. Besides the explanations given in the Woodruff articles, a good explanation for the true cause of the rotation of the radiometer appears in Gilbert D. West, "A Modified Theory of the Crookes Radiometer," *Proceedings of the Physical Society of London* 32: 222–231 (December 1919), which revives Sutherland's work from *Philosophical Magazine* 42: 373ff. and 476ff. (1896).

8. P. Lebedew, "Experimental Investigation of the Pressure of Light," *The Astrophysical Journal* 15: 60 (1902); Peter Lebedew, "An Experimental Investigation of the Pressure of Light," *Annual Report of the Board of Regents of the Smithsonian Institution* (1903), 177–178; P. Lebedev, "Untersuchungen über die Druckkräfte des Lichtes," *Annalen der Physik* 311 (11): 433–458 (1901).

9. E. F. Nichols and G. F. Hull, "The Pressure due to Radiation," *The Astrophysical Journal* 17 (5): 315–351 (1903), online at http://books.google.com/books?id=8n8OAAAAIAAJ&pg=RA5-PA327&dq=torsion+balance+radiation#v=onepage&q&f=false, accessed October 7, 2012.

10. R. A. Beth, "Mechanical Detection and Measurement of the Angular Momentum of Light," *Phys. Rev.* 50: 115–125 (1936).

11. Lynn Yaris, "Nano Sized Light Mill Drives Micro-Sized Disc," *Physics.org* (July 5, 2010), online at http://phys.org/news197555841.html, accessed October 7, 2012. The original article is Ming Liu, Thomas Zentgraf, Yongmin Liu, Guy Bartal, and Xiang Zheng "Light-Driven Nanoscale Plasmonic Motors," *Nature Nanotechnology* **5** (8): 570–573 (2010), available online at http://xlab.me.berkeley.edu/publications../pdfs/142.NatNanotech2010_Ming.pdf, accessed October 7, 2012.

12. B. Denardo and T. G. Simmons, "An Acoustic Radiometer," *Am. J. Phys.* 72: 843–845 (June 2004).

13. See reference 1.

{ 18 }

Sound Movies, the World's Fair, and Stellar Spectroscopy

The History of Ideas shows peculiar gaps, in which it seems as if no progress is made for long periods of time. Often, when you examine these gaps, you find out why those gaps are there—some critical piece of equipment was missing, perhaps, or some philosophical jump was needed. Then again, sometimes it's not at all clear why no progress was made. The case of Stellar Spectroscopy has both sorts of gaps. Along the way, we'll also have a look at the gap between the movies and sound movies, and the interesting opening of the 1933 Century of Progress World's Fair in Chicago. It will all come together, I assure you.

In a very rough way, of course, stellar spectroscopy has an ancient history—the observation that different stars are different colors goes back to antiquity, with very clear writings by the Greek philosophers on the subject, and some possible observations recorded in myths and folklore. But the scientific observation of the spectra of astronomical bodies clearly took a giant step forward when Isaac Newton used his prism to break up light from the sun into its constituent colors, and to reassemble them into white light. Newton himself does not appear to have used his prism to study starlight, nor do any of those who repeated his experiments. At any rate, if they did so, they don't seem to have recorded it.

This isn't altogether surprising—starlight is much dimmer than sunlight. What is surprising is that no one seems to have addressed the issue for the century between Newton's *Opticks* and the beginning of the nineteenth century. It would have been relatively simple to look at a star through a telescope with a prism near the eyepiece, however, and observe the color separation. Nevertheless, no one seems to have done this until Joseph Fraunhofer in 1814. Fraunhofer had already repeated William Wollaston's 1802 experiment of observing the solar spectrum and noting the dark absorption bands. But Fraunhofer carried his work on the solar spectrum past qualitative observation to quantitative measurement. He constructed a diffraction grating by winding wire on a frame, and used the known spacing to determine the wavelengths he was observing. After he did this with the solar spectrum, observing, measuring, and naming what are now call the *Fraunhofer lines*, he did the same with stellar spectra. He noted similarities between the solar spectrum and that of Betelgeuse, and differences with that of Sirius. It's not surprising that he concentrated on the brightest stars at first—they would be by far the easiest ones to observe. He would have undoubtedly moved on to other stars, had he not died of tuberculosis at the relatively young age of 39 in 1826.[1]

Perhaps it was Fraunhofer's amazing early use of a grating to calibrate wavelength that is responsible for his work being recorded and transmitted. Wollaston's observation of the dark bands resulted from his using a prism with greater resolving power than had been used before. Fraunhofer's contribution was the measurement of wavelengths. After his death, others carried on the work on the solar spectrum, refining the measurements. But his work on stellar spectra lay pretty much ignored for about 40 years.

Why? One wonders. Certainly there was no shortage of stars to study, and there were many discoveries yet to make in the study of stellar spectra. I thought that the answer might be the lack of a good recording medium. Fraunhofer did his work entirely by eye. Photography—especially technical photography for scientific measurement—was in its infancy, so the first solar spectra were drawn by hand. Edmond Becquerel photographed the solar spectrum in 1842, and J. W. Draper repeated this the next year. But stellar spectra, of course, are significantly dimmer. In the 1860s such pioneers as G. B. Donati and Angelo Secchi drew their stellar spectra by hand, as Fraunhofer had. So the lack of recording media does not explain the lacuna in stellar spectroscopy.

Nevertheless, the 1860s saw a sudden blossoming of interest in stellar spectra, with independent and early simultaneous work by Donati and Secchi in Italy, by Royal Astronomer George Biddell Airy at Greenwich, by Lewis Rutherford in New York, and by William Huggins and William Allen Miller in London. It soon became clear that there were certain characteristic sets of absorption lines, and that stars tended to fall into a few definite types.

The next step beyond this was to record not only the wavelengths of these lines but also their relative widths and strengths. Even with photography as a tool, these parameters were not so easy to establish. What was needed was something that could measure the strength of the signal and record it. There had been efforts to build photometers dating back to John Herschel in the 1830s. But these depended upon human eyes to compare the intensity of a light to some reference standard. Some sort of objective device, not dependent upon human judgment, was needed.

The breakthrough for this came in 1873 when Willoughby Smith discovered photoconductivity in Selenium. Alexander Graham Bell used this effect to create his *photophone* in 1880, an invention that he considered the greatest he ever made, surpassing the telephone. It used a beam of light to transmit telephone conversations across distances without any wires between the sender and the receiver. A year later the selenium cell was used to produce what was essentially an early FAX machine, capable of sending pictures over telephone lines.

With the selenium photoconductor and the selenium photocell one could measure the relative strength of a signal automatically and objectively. But it still couldn't be recorded except perhaps in the most delicate ways—using a whisker, perhaps, to make a track in a carbon film on smoked glass. In order to do something useful, such as using a motor to drive a pens to make a permanent record, would require the signal to be amplified. And there was, as yet, no way to do this. Until signals could be amplified, selenium cells (and the other photocells being discovered) had to remain limited devices, mainly of scientific value and novelty items.

The lack of a good amplifier for electrical circuits made itself felt in other areas as well. Consider the motion picture. Louis le Prince, arguably, had invented it in 1888. Most sources will tell you that the first sound movie didn't appear until 1927, when *The Jazz Singer* was released. A few will admit that there were less well-known experiments prior to this, but it is true that commercial sound films effectively date from *The Jazz Singer*'s release. But the sound in this film came from what were, in essence, phonograph records—why did this take so long to develop? Edison is responsible for both the phonograph and for helping to develop the motion picture. Is it likely that he wouldn't connect the two? We know, in fact, that he did. The Dickson Experimental Sound Film showing a man (William Dickson, who made the film in collaboration with Edison) playing a violin into an oversized phonograph bell while two men dance. Clearly there was a sound recording to go with that film footage. The accompanying sound recording has been found, and in 2000 was restored and synchronized with the film.[2] The recording was made in 1894 or 1895, and there were, in fact, many record-accompanied films after that date. So why do they count the relatively late *Jazz Singer* as the first sound film?

The answer is that the effective audience for such a sound film was limited by the range of the phonograph providing the sound. And, at first, the sound was generated by a needle vibrating in a groove, amplified by the effect of the metal "bell" attached to the diaphragm. Such a mechanical amplifier could not provide much range, limiting the prospective audience to a small roomful of people. The missing item needed to make sound movies practical for a real audience was a reliable electrical amplifier.

The needed element was the thermionic valve, colloquially known as the *vacuum tube*, invented in 1906 but not effectively used until 1916–18. It was almost immediately used in radios, boosting the rectified signal from a tiny output requiring earphones to a healthy output that could drive a large speaker. Radio could go from a single-listener device to a device that could play to an entire room. Broadcasting went from a hobby to an industry. More powerful amplifiers allowed sound from a recording to fill an auditorium and made sound movies a commercial reality. Amplifiers were used to boost telephone signals.

In the early 1930s vacuum tubes allowed electronics to boost the feeble output of photocells to power motors as well, and to give the oomph needed to trip relays.

A burst of engineering creativity provided Electric Eyes that opened doors and moved things for those crippled by polio.[3,4,5]

The Century of Progress World's Fair was set to open in 1933 in Chicago, celebrating the city's centennial. The previous Fair held in Chicago had been 40 years earlier in 1893. Someone realized that it would be a coup if they could somehow tie the new Fair to that older one in a technological way—the theme of the exhibition was, after all, the Century of Progress. Noting that the star Arcturus was 37 light-years away, someone saw that the light just arriving at Earth would have left shortly after the time of the previous exhibition. It as decided to turn the lights on for the Fair for the first time using the light from Arcturus. With the new vacuum tube amplifiers, the signal from a photocell placed in a concentrator at the end of a telescope could be amplified enough to trip a set of relays. So on May 27, 1933, light from Arcturus started a Rube Goldbergian chain of events that led to the powering of the lights at the Chicago World's Fair.[6]

One problem with these early amplifiers was the lack of linearity. Harold Stephen Black, then working for Bell Labs, discovered a way to improve the linearity of signals by using feedback from the amplified voltage back into the circuit, something that came to be called *negative* feedback. He applied for a patent, but it took almost a decade to be granted. As Black later wrote,[7] "One reason for the delay was that the concept was so contrary to established beliefs that the Patent Office initially did not believe it would work." With this sort of control, amplifiers for recording of data, which had been used since the invention of the triode, could now be made more reliably, and became a standard piece of laboratory apparatus from the 1930s on. Chart recorders that used amplified signals from photodetectors to take spectra. Scientific spectroscopy had advanced from a labor-intensive art to a laboratory tool.

Notes

1. J. B. Hearnshaw, *The Analysis of Starlight: One Hundred and Fifty Years of Astronomical Spectroscopy* (1986).

2. You can see and hear it at http://archive.org/details/dicksonfilmtwo, accessed August 9, 2012.

3. W. C. Douglas and O. H. Caldwell, "The Electric Eye," *The Polio Chronicle* (October 1932).

4. "The Latest Scientific Marvel: An Electric Eye," *New York Times Sunday Magazine*, December 23, 1906, SM2.

5. "Electric Eye Burglar Alarm," *Argus* (Melbourne, Australia)(September 16, 1922), 8.

6. Cheryl Ganz, *The 1933 Chicago World's Fair: A Century of Progress* (2008).

7. Harold Stephen Black, "Inventing the Negative Feedback Amplifier," *IEEE Spectrum* (December 1977), cited in http://en.wikipedia.org/wiki/Negative_feedback_amplifier, accessed August 9, 2012.

{ 19 }

Déjà Vu

In "Matter of Fact," the fourth installment of his television series *The Day the Universe Changed*, James Burke makes the intriguing argument that one of the unexpected results of the invention of printing with movable type was the rise of indexing. More than any prior development, this provided much greater access to information becoming available through the new printed books, which had previously been available only to the few rare widely read individuals who had waded through the collections of poorly labeled scrolls and omnibus bound volumes. Now that people had a better idea what had been printed, through card catalogs, lists of printed books, and indices of those books themselves, information that had been written down but buried within a few badly known volumes became free to a wider range of interested readers and researchers.

Through the years more and better indices and databases became available. When I was working on my undergraduate degree, I could search for relevant scientific papers using the *Physics Abstracts*, the *Science Citation Index*, and the *Cumulated Index Medicus* (which had surprising references to physics topics buried in it), as well as the indexes published by individual magazines and unlikely sources such as the *New York Times Index*, the *Reader's Guide to Periodic Literature*, and the *19th Century Guide to Periodic Literature*. Dogged searching, cross-referencing, and following links backwards to the source could all be used to ferret out almost all the relevant papers and references on a topic.

Today, of course, databases are easily accessible and searchable through computers. Who needs the tedium of going through the hard-copy published version of the *Science Citation Index* when you can search it using the online database *Web of Science/Science Citation Index*? Even centuries-old issues of the *Journal of the Royal Academy of Sciences of London* can be searched and downloaded through the *JSTOR* database. Arguably of even greater importance is the development of the Search Engine, which serves to link items not linked by their creators. The services that *Google* has instituted—*Google Scholar*, *Google Books*, the *Google Books Ngram* search, among others—provide yet other methods of performing searches through literature both electronic and hard copy. But I'll return to this later.

What I want to discuss here are those cases of discoveries and ideas which somehow or other failed to be caught by the pre-computerized searches, either because they were not well represented in the indices, or because those doing the searches weren't diligent enough. When I was researching my book on mythology, *Medusa*, I was surprised to find that the idea that the Gorgon's snaky head might have been inspired by the body of an octopus, squid, or cuttlefish had been proposed not merely once or twice, but a dozen times between the end of the nineteenth century and the end of the twentieth, and that none of these authors seemed to be aware of those who had previously suggested the idea (with the exception of one author who discovered a prior article on the topic, but only just before publication).

How can this be? Surely anyone seeking to publish an idea will search the prior literature. And it is true that I only learned of three of those suggesting the idea through footnotes in more comprehensive surveys. The original articles were either in very obscure journals, or in another language, or were not properly published, simply orally communicated. But in other cases the articles had been prominently printed in notable journals that were themselves regularly indexed—*Folk-Lore*, for instance, or the Freudian journal *Imago* (in which the idea was proposed twice!). The truth seems to be that people simply didn't search carefully enough.

Well, that's all in the Humanities, you might argue. The indices and databases in the Humanities used to be less well-developed and interconnected. It would have been easy to miss some references. But surely in the hard sciences the situation would be different. Not necessarily. One piece of information I used in my doctoral thesis was the distribution of closest neighbors. Let's call it **P(r)**. Given some items scattered at random throughout a volume with average density $\boldsymbol{\rho}$, the probability that the closest of these items to a specific point is a distance **r** away is **P(r)**.

One would expect, on the basis of dimensional analysis alone, that the most likely separation is on the order of $\mathbf{1/\rho^{1/3}}$. That's also the result you get by assuming the entities are equally spaced on the points of a simple cubic lattice. But this doesn't really tell us the distribution. I needed the form of the distribution to calculate its moments. From time to time, other people have needed it, too. I was very surprised to find that some of these previous workers were evidently unaware of prior work—they didn't cite any, and one of them seemed a little unsure of the result. Yet the topic has a history going back to 1909. There is even a historical review of the topic.[1] The most elegant derivation of the result is that by astrophysicist Subrahmanyan Chandrasekhar (who was using it to determine how close together galaxies were likely to be). It takes only two lines.[2] For those interested, the probability distribution **P(r)** that the nearest neighbor lies between **r** and **r + dr** is

$$P(r)dr = 4r^2\pi\rho\exp\left\{-\frac{4}{3}\pi r^3\rho\right\}dr$$

This function peaks at $\mathbf{r_{max} = (2\pi\rho)^{-1/3}}$, and the most likely value is $\mathbf{0.554\rho^{-1/3}}$, about half the value one would naively expect, based on that simple cubic model.

Well, okay, this is just a relatively unimportant footnote to items of greater importance—people can be forgiven for overlooking previous efforts. They looked for the distribution themselves, performed a few tests to convince themselves that they had it right, then moved on. You can't spend a lot of time on *every* detail, and they did all end up with the correct distribution function.

But what about a new development in physics? A new type of laser medium, say. Surely people would notice that, wouldn't they? Again, not necessarily. I was very surprised, when I started working with color center lasers and researching their history. Color centers are point defects in insulating crystals. In many of these, particularly the ones in alkali halides like common salt, the defect has a big absorption in the visible regime and will turn such crystals from colorless to brilliantly colored. I learned that color center expert J. J. Markham had suggested using color centers as a laser gain medium even before the first successful laser had been built. He was, of course, familiar with the physics and spectroscopy of these point defects in ionic crystals and knew that the F-Center in potassium chloride constituted an ideal four-level system that could be pumped very hard without fear of saturation. (It's somewhat ironic that the first physically realized laser was the three-level ruby laser.) Some Russian journals suggested the use of color centers as laser sources in the early 1960s.

Thereafter, the idea came up repeatedly as a likely source of lasing. I have seen it suggested in two cases where the experimenters claimed to have measured gain using the F-center in KCl. They either did not try to build the laser themselves, or did not report any attempts they made. The modern history of color center lasers effectively begins with Mollenauer and Olson's[3] 1975 paper "Broadly Tunable Lasers Using Color Centers." But this was not the first paper reporting successful lasing of a color center, as the authors themselves acknowledged by citing the first one.

In 1965 Berndt Fritz and E. Menke had succeeded in lasing a crystal of lithium-doped potassium chloride containing $\mathbf{F_A(II)}$ centers (an F-center perturbed by an adjacent foreign Lithium ion). The crystal was cooled to liquid nitrogen temperature and pumped—surprisingly—by a flash lamp, the same as a ruby laser (almost all early experiments with color center lasers afterwards used another laser as a pump source, mostly because of the very rapid fluorescence decay rate). This was reported in *Solid State Communications* in 1965.[4] What is surprising is that this appeared years before many of the proposals that F-centers might make good lasers were written. The authors of these proposals were unaware that such a laser had actually been built and operated.

The article was reported in the usual indices and collections of abstracts. But that it was overlooked is more surprising because there were few comprehensive books on color centers at the time. One of the most prominent was *Physics of Color Centers*, edited by W. B. Fowler and published in 1968 by Academic Press. The book reports on Fritz and Menke's laser and includes it in the index as the sole entry under "lasers," which would be hard to miss.

I suspect that the explanation, again, is that these suggestions were not the main focus of the works that made them. The suggestion is that this material, among others, might be a good laser source, or that, gain having been measured in this material, the next obvious step is to build a laser using it. But the main thrust in all cases was elsewhere, with the use of color centers in lasers as an almost offhand suggestion that didn't require closer scrutiny or research. I know from experience that it's such apparently unimportant side comments and footnotes that come back to bite one later. Since they're not checked, it's easy to miss things or to make errors. The moral of the story is to take the time to carefully examine even what you think are obvious or unimportant throwaway remarks.

Which brings us, finally, to the promised topic of computer search engines. Such items, which appear to be so efficient and useful, can give a false sense of security. In the first place, they're not so efficient—I only have to perform a "vanity search" on myself in the databases or in common search engines to find that an awful lot of my own list of publications hasn't made their lists. But even if (when?) such databases and search engines become more dependable, they still cannot automatically correct for the items and topics that the researcher and author does not trouble him- or herself to examine in more detail. In other words, it's not the computer's fault if you fail to do your research in sufficient detail.

Notes

1. M. B. Santos, "Distribution of Neighbors Other Than the Nearest," *American Journal of Physics* 54 (12): 1139–1141 (1986), and addendum *Am J Phys* 55: 952 (1987).

2. Subrahmanyan Chandrasekhar, "Stochastic Problems in Physics and Astronomy," *Review of Modern Physics* 15 (1): 1–89 (1943).

3. L. F. Mollenauer and D. H. Olson, "Broadly Tunable Lasers Using Color Centers," *Applied Physics Letters* 24 (8): 386 (1974). Also in *J. Applied Physics* 46 (7): 3109–3118 (1975) and *IEEE J. Quantum Electronics* 10 (9): 712 (1874).

4. B. Fritz and E. Mencke, "Laser Effect in KCl:Li with FA(II) Centers," *Solid State Communications* 3 (3): 61–63 (1965).

{ 20 }

The Magic Lantern of Omar Khayyam

For in and out, above, about, below,
'Tis nothing but a Magic Shadow-show,
Play'd in a Box whose Candle is the Sun,
Round which we Phantom Figures come and go.[1]

Although three volumes that called themselves Mark Twain's *Autobiography* were published before 2010, Twain himself stipulated that his *real* autobiography could not be published until a century after his death. This guaranteed that no one living at the time might still be around to be offended. The first volume, published in 2010, contained much that had never seen print before. One of these items was the revelation that he originally intended to use as an epigraph the excerpt from Edward Fitzgerald's translation of the *Rubaiyat* of Omar Khayyam reproduced above. Twain admired it, saying "No poem had ever given me so much pleasure before, and none has given me so much pleasure since; it is the only poem I have ever carried about with me; it has not been from under my hand for twenty-eight years." Khayyam's work was immensely popular at the time, capturing the fatalist spirit of the late nineteenth century. The quoted verse describes how we humans are like puppets in a Magic Shadow Show. But what exactly does that mean?

The wording suggests the Magic Lantern Show, a public exhibition of transparent slides projected onto a screen using an imaging lens. Certainly that is how at least one Internet site and two books interpret it.[2] In addition, Edward Byles Cowell, Fitzgerald's friend (and his tutor in Farsi) published his own translation, in which he renders this line as "the image of a magic lanthorn."

But Omar Khayyam was born in 1048 and died in 1131. Does the art and science of projecting images date back that far? Or did Khayyam—a noted mathematician as well as a poet—anticipate the invention?

Many inventors and dates have been suggested for the Magic Lantern, but it is generally now acknowledged that the inventors were Athanasius Kircher, who demonstrated the principle in 1646, and Christiaan Huygens, the champion of the wave theory of light. He had produced one by 1659. Curiously, he seems to have been ashamed of this invention, although it was evidently a popular novelty.

Shortly after Huygens introduced it, others began using it and putting on exhibitions. One of these, Thomas Walgensten, coined the term "Magic Lantern" (*Laterna Magica*), by which it has been known ever since. The devices were used in magic shows and early "spook house" exhibitions called *Phantasmagoriae* from the eighteenth century onward. (Jules Verne describes such a use in his 1893 novel *Le Château des Carpathes* [Carpathian Castle]) The heyday of the Magic Lantern was the second half of the nineteenth century, when traveling exhibitors took their lanterns and stacks of glass slides on the road, showing photographs of far away places and news events. They also had brightly colored drawings with which they told stories, and even "animated" slides with moving parts so that they could show moving figures.

This was when Fitzgerald composed his translation of the *Rubaiyat*, and it was perhaps inevitable that suggestions of this very lively image—God as Heavenly proto-cinema director—appeared in his work. But looking at it closely he doesn't *actually* call it a Magic Lantern show. The words *Magic Shadow-show* might be seen as suggesting it, but they might also mean a shadow play, using shadows generated by figures or hand gestures to tell a story.

What does Khayyam's verse actually say? Transliterated from the Farsi, it reads[3]:

> În charkh-I falak ki mā dar ū hairān-ūn
> fānūs-I khayāl az ū mithālĭ dārūm
> khwurshĭd.chirāgh.-dānu 'ālam fānūs
> mā chūn ṣuwar-ĭm.k-andar ū gardān-ĭm

Peter Avery and John Heath-Stubbs' 1981 translation renders this as:

> Let us consider this wheel of heaven that amazes us,
> As if it were a diorama—
> The sun the candle, the world the lanthorn,
> Then we are like the images revolving on its walls.

The use of "diorama" seems an unwarranted modernism as well and can refer to either a nineteenth-century popular entertainment or a modern museum setting. Another uncredited translation renders it "Magic Shadow-Shapes." The implication seems to be that reality is compared to some fixed mechanical device that throws shadows on a wall far away.

Such images have a long history themselves. If they are not as detailed and complex as Magic Lantern images, they are easier to produce, requiring no lenses or mirrors. The punched-tin lanterns used in the American Colonial era, used when the transparent windows of glass or mica were too breakable or too dear, projected pinhole images. You can see an example in the fifth painting (or engraving) of William Hogarth's 1743 series, *Marriage à-la-Mode*, where the Night Watch's lantern projects such spots on the ceiling. Similar devices had been used much earlier. The circa 1420 book *Liber Instrumentorum* of Giovanni de Fontana has an illustration showing a man holding a lantern, and beyond him is an image of a devil projected on a wall.[4] The principle probably goes back much farther than this. Many ancient Byzantine, Greek, and Roman oil lamps have handles that make carrying convenient, but which stick up so far that the handle itself must cast a shadow from the flame. Most handles are kept low, apparently to avoid casting such an undesirable shadow. The ones that do cast shadows are notably in distinctive shapes—crosses, crescent moons, and the like. Some lamps even have small statues mounted on them that must necessarily have cast human-shaped shadows.

Of course, mention of shadows brings up the idea of hand shadows and shadow plays, such as the famous shadow-plays of Burma, made by casting shadows from stylized metal puppets onto a screen. This type of entertainment has a long history as well, being first described in China during the Han dynasty (circa 200 B.C.E.–200 C.E.). It has been suggested, on oblique evidence, that it may date back to Harappan times (3000–1500 B.C.E) in India. Exactly when such plays were first performed elsewhere is still obscure, but there is a report that Saladin, who died in 1193, saw such a play, which puts it within reach of Khayyam's lifetime.[5]

There is an early suggestion of shadow plays in the West in Plato's famous Allegory of the Cave, given in Book 7 of *The Republic*. In this, he describes men chained in a cave, unable to turn their heads, but only able to see shadows projected onto the wall in front of them by puppeteers, and which they take to be Reality. His allegory, in fact, recalls Omar Khayyam's imagery that started this essay—what if what we call Reality is simply the shadows we see on the wall of a cave underground? It's true that Plato never compares this conception to a popular or religious play, but it would not be unknown for a Greek description not to invoke its inspiration directly. And the description of The Cave in *The Republic* seems very much like a performance, right down to actors providing voices for the shadow figures. It's easier to believe that Plato simply recruited the familiar image of a popular shadow-play into service to make a philosophical point than that he constructed an elaborate, complex, and artificial situation to make his point.

As for hand-shadows, they were a popular Victorian amusement, but there don't seem to be any earlier descriptions of them. Nevertheless, it seems equally hard to believe that they did not have a very long prior history.

In the end, it seems probable that Omar Khayyam's verses describe our world as an ephemeral construction as unreal in essence as the shadows thrown by a shadow-lantern. The relevant words are fānūs-I khayāl, and translate approximately as "lantern of imagination," with "imagination" being a philosophical term derived from Aristotelian philosophy. Fitzgerald's friend Cowell himself claimed that this verse referred to "A magic lanthorn still used in India; the cylindrical interior being painted with various figures, and so lightly poised and ventilated as to revolve round the candle lighted within."[6] Fitzgerald's original translation of the verse,[7] prior to the first published edition of the *Rubaiyat*, rendered this as "Visionary Shapes," which seems to indicate that he had Plato's Cave in mind, after all. But it's obscure to the new reader, and doesn't flow poetically, so we have the more beautiful but less accurate translations as "lanterns" and "shadow plays" in published verse. A present-day poet might describe Reality as no more real than a movie, or in more up-to-date terms as a video game, or a virtual reality experience. Our metaphors are determined by our culture and experience.

Notes

I'd like to thank Historian of Science Elaheh Kheirandish for many helpful discussions regarding the correct translation of the verses. I hasten to point out that any errors are mine.

1. This is the version of the quatrain from the fifth edition of Fitzgerald's translation of the *Rubaiyat*, probably the most familiar and quoted. Twain's use of the poem as an intended epigraph is noted in *Autobiography of Mark Twain*, Vol. 1, ed. Harriet Elinor Smith (2010), 525. A reproduction of the page with the quotation appears on page 14 of that volume, which says that it forms the first page of the manuscript "My Autobiography [Random Extracts from It]." It also notes that Twain deleted it from later copies of his autobiography, and so the stanza only shows up as a footnote. Besides Mark Twain, pulp author Robert E. Howard (creator of *Conan the Barbarian*) used this verse as the first chapter epigraph for his 1929 novella *Skull Face*, his answer to Sax Rohmer's *Fu Manchu*.

2. http://billdouglas.ex.ac.uk/eve/exhibitions/exhibitions_print.asp?esID=305, accessed September 22, 2012, and Vachel Lindsay, *The Art of the Moving Picture* (1915) and William J. Elliott, *Shadow Show* (1942).

3. From A. J. Arberry, *The Romance of the Rubáiyát: Edward Fitzgerald's First Edition* (1959), 220.

4. *A History of the Magic Lantern*, page 3 at http://www.magiclantern.org.uk/history/history3.html, accessed September 22, 2012.

5. Paul Kahle, "The Arabic Shadow Play in Egypt," *J. Royal Asiatic Society of Great Britain and Ireland* **1** (January 1940); and Fan Pen Chen, "Shadow Theaters of the World," *Asian Folklore Studies* **62** (1): 25–64 (2003). Fan Pen Chen seems highly skeptical of many of the Shadow-Play traditions. He doesn't seem to think Plato's allegory of The Cave is linked to a tradition of performing shadow plays. He cites the verse from the *Rubaiyat*, but doesn't name Omar Khayyam as the author, and he thinks the verses describe "an example of the zoetrope type of lantern in which shadow figures gyrate around a lamp," much as Cowell did.

6. Edward Byles Cowell, "Omar Khayyam," *Calcutta Review*, March 1858.

7. For further information, see Arberry, *The Romance of the Rubaiyat*; and *The Rubaiyat of Omar Khayyam*, trans. Robert Graves and Omar Ali-Shah (1968).

PART II

Weird Science

{ 21 }

The Yellow Sun Paradox

Though I am old with wandering
Through hollow lands and hilly lands,
I will find out where she has gone,
And kiss her lips and take her hands;
And walk among long dappled grass,
And pluck till time and times are done
The silver apples of the moon,
The golden apples of the sun.

—THE LAST VERSE OF *THE SONG OF WANDERING AENGUS* BY WILLIAM BUTLER YEATS (1865–1939)[1]

If you ask a preschooler to draw a picture of the sun, they'll draw it as a yellow circle, often with visible rays coming from it (and maybe a smiling face). The Gold Ball in the story of The Princess and the Frog represents the Sun, said mythologist Joseph Campbell, because Gold is the Solar Metal. Egyptian, Celtic, Chinese, and Aztec representations of the sun are made of gold. If you ask the average person on the street (but not an astronomer or astrophysicist) what color the sun is, they'd say it was yellow. This is common knowledge and common experience.

And yet the sun is *NOT* yellow. Sunlight is the very definition of white light, quite literally. If the sun were truly yellow, then the colors of everything we see would be subtly altered. White signs, white houses, white sheets drying on the line seen on a bright sunny day would appear to be yellow. Working under truly yellow light—as in a lithographic facility—can be unnerving. The CIE coordinates of the standard illuminants—the measurements and definitions used by color experts to quantify color—all lie close to **(0.3,0.3)**, the white locus of the Color Diagram.

The sun is undoubtedly white, yet everyone seems to perceive the sun as yellow. These statements cannot possibly both be correct. Addressing the problem, Phil Plait, who has published books and runs a website as "The Bad Astronomer" (meaning one who exposes Bad Astronomy in popular culture and the like, not an astronomer who is somehow bad), remarked that "It's not really well-known why [the sun is thought to be yellow]."[2] He then goes on to give some reasons proposed for why the sun is yellow.

One is that the same Rayleigh scattering that is responsible for the sky being blue also makes the sun appear yellow, since some of the blue has been scattered out. (This is the most common suggestion I hear when I mention the paradox to people.) But the amount of blue light scattered out is far too small to have a noticeable effect on the color of the sun. The CIE standard illuminants like **B**, **C**, and **D65** already have the effects of scattering built into them, and they predict a very white sun.[3]

A second suggestion is that we perceive the sun as yellow because we are comparing it to a blue sky. Perception studies show that the background can, indeed, affect the color we seem to see. But, as Plait cleverly points out, if white objects appeared yellow by comparison with the blue sky, then clouds would seem to be yellow, and they're not.

A third possibility is that we cannot see the sun when it's high in the sky because it's too bright and looking at it hurts your eyes. When the sun gets low, closer to the horizon, we can look at it, because more of its light gets scattered away by the atmosphere. Perhaps not so coincidentally, this is when the sun's color changes, precisely because of that scattering. So it appears yellowish, then orange, then red, and finally magenta. To quote Plait, "Since it looks yellowish at the only time we can really see it, we remember it that way. This is an interesting claim, but there are some problems. I remember it most when the sun is a glowing orange or red ember on the horizon, and not yellow, so why don't I think the Sun is red?"

A fourth explanation I have found on the Internet is that the sun appears yellow because "that is the way the brain processes the data from the eye," as if that somehow explains it. But this is simply passing the buck from our ocular system to the cerebral cortex and doesn't really explain anything. Why doesn't the brain process those clouds and white sheets as yellow? The wavelength makeup is the same, whether light comes from the sun or the cloud.

So, once again, why IS the sun yellow in our thoughts? We can calculate the color of something by starting with the wavelength distribution of the light and performing some simple calculations using scattering formulae that model how the light is attenuated as it passes through the atmosphere. The results give a locus of points on the chromaticity diagram that can be uniquely identified with the colors produced. [4]

I decided to model the passage of the light from the sun through an atmosphere that scattered according to a strict Rayleigh scatter $\mathbf{1/\lambda^4}$ law. I assumed Illuminant **D65** (Noon Daylight), a scattering cross-section that depends upon the inverse fourth power of the wavelength (and a loss exponential in the product of this cross-section times the optical path length, times a constant), and the usual three standard color functions. I then calculated the CIE Chromaticity coordinates **(x,y)** in the usual fashion by numerical integration of the product of the illuminant, scattering function, and color over wavelength space, then normalizing the chromaticity coordinates **X, Y,** and **Z**. Using the same procedure, and looking at the color of the light thus scattered *out* of the incoming sunlight, you can predict that the sky should be blue. In this case, I looked at the color of the sun as the light as scattered out.

The results were very interesting. The starting point, with negligible scatter, was the illuminant **D65** "white point" of **(0.313, 0.329)**, but as soon as the path lengthened, the trajectory of the locus of the apparent sun color started moving directly toward the Spectral Locus at about 570 nm, which is about as yellow as you can get. It continued toward this point for some time before veering off slowly toward about 580 nm, which is still well within what is generally termed "yellow," then gradually turning orange and then red, and asymptotically approaching the deep red terminus of the Spectral Locus.

Its trajectory superficially resembles the Planckian Locus, representing the perceived color of a Blackbody Radiator as it cools, but the differences are significant. The Blackbody starts not at the White center, but at the limiting point of **(0.328, 0.502)**, at the light-blue color of Blue Heat. It then arcs across, skirting the edge of the White region at about 6000K before cutting across the yellow portion of the spectrum between about 4000K and 2500K before asymptotically approaching the red end of the Spectral Locus about the same way the sun color curve does. The difference is that the Planckian Locus curve starts in the blue and spends much less of its length in the yellow portion of the color diagram.

It's not just coincidental that the sun starts to seem yellow and continues to be yellow for some time after the light begins to attenuate—this yellow is the complement of the blue sky, and on the Chromaticity Diagram is diametrically opposite the Blue Sky locus (which this calculation sets at **(0.2279, 0.2312)** in the limit of small amounts of scatter). The chromaticity coordinates of the blue sky changes very slowly with increased scatter distance (ultimately moving towards the White locus as the scattering length approaches infinity), and when you subtract this blue from the white you get the yellow as residue. In essence, the sun appears yellow because of the blue scattered out of it, much as people surmised in the first option above, but not until enough light is scattered out to make the sun directly viewable. So why do I say that that first explanation isn't quite right?

Because the sun at its zenith doesn't have enough blue scattered out to look yellow. But as it descends more and more blue gets scattered out, and it gradually turns more yellowish as it descends. Until the sun gets very low in the sky and starts changing to orange and then to red, it spends all of its time either as white or as yellow. As soon as it is attenuated enough to look at it even fleetingly, it appears to be yellow, and continues this way until it very clearly and rapidly begins to change color at sunrise or sunset. So the "unadulterated" color of the sun is yellow. Thus, we know that very close to sunrise or sunset the sun often turns orange or red, but we recognize that this is not its usual color. During those times when we can get a fleeting glimpse of it, its color is white or yellow-white, and it often appears yellow when the sun has risen above the horizon, or is about to go there. At such times white objects directly illuminated by the still-attenuated sun do indeed have a golden color to them. At the same time, white clouds far overhead will appear to be white, not yellow, because they are illuminated by rays that have not passed through as much attenuating atmosphere.

When the sun is attenuated behind clouds, so that you can look at it without being dazzled, it appears white or yellowish. Thus, we get the impression that, while the sun may turn reddish at sunset, yellow, often yellow too bright to look at (so it appears white) is the sun's "true" color.

I've deliberately made the observations on clear mornings when I can see white objects appearing to the yellow in the direct rays of the low sun, but behind them I can see the clouds as white, with no hint of yellow. Some people have suggested that sunlight is yellow because of the blue being attenuated out of it, but that objects that are white appear to be white because they get the direct yellow sunlight, but they also get the blue light coming from all parts of the sky. This, they say, explains white clouds. But it doesn't explain the yellow-appearing church and white fence closer to me, which is illuminated by the same sky. Another counterexample is the fact that my white porcelain bathroom fixtures appear white when illuminated by sunlight from a sun high in the sky passing through my bathroom window, even though the walls of my house are excluding most of that blue contribution from the rest of the sky.

But this isn't an end to it. When you look at the sun, you don't simply see a white circle (or a yellow, orange, or red one, depending on how low in the sky it is) in a blue sky. If you block out the sun with a nearby object, or even cover it with your thumb, you can clearly see that there is a glare surrounding the sun. This is due to forward-scattered light, and it can be quite bright. If you remove the obscuration and look briefly but directly at the sun you will see a much brighter and more encompassing glow surrounding the sun, due to the entoptic scattering within your eye. There are many contributors to this inside the eye scatter—scatter from the corneal tissues, the crystalline lens, and so on. Much of the scattering, both inside and outside the eye, is yellow in color, even when the sun is high in the sky and should, by the above arguments, appear white. Again, you cannot look comfortably at the sun itself, but this yellow halo of glare certainly conveys the impression that the too-bright-to-look-at sun itself is a very intense yellow.

There is also another possible contribution to the idea of a yellow sun. Early people naturally viewed the sun as a sort of flame, since such combustion was their daily experience—the sun was a large fire in the sky. One would therefore expect the sun to be the same color as flame.[5]

Exactly what color flame is depends upon what you are burning, and what part of the flame you're looking at, but these are academic quibbles. The ancient Egyptians, Babylonians, Greeks, and Chinese weren't burning propane with a blue flame—they were looking at wood fires and wax candles and oil lamp flames, and would surely be concentrating on that part of the flame which takes up the largest portion, yielding the most light—the broad yellow expanse of the flame that consists of heated particles of soot heated up by combustion and glowing with blackbody radiation. That would be roughly 1900K blackbody radiation, corresponding to a spectral color of about 590 nm, about at the juncture of the yellow and orange regions. Early people, expecting the sun to be a flame, would expect it to be yellow-orange, and experience would confirm that it was yellow, as soon as it was dim enough to be seen directly.

Notes

1. *The Collected Poems of W. B. Yeats*, ed. Richard J. Finneran (1987).

2. The Bad Astronomer's discussion of the Yellow Sun is on pages 45–47 of Philip C. Plait, *Bad Astronomy: Misconceptions and Misuses Revealed, from Astrology to the Moon Landing "Hoax"* (2002). Also see his website at http://blogs.discovermagazine.com/badastronomy/, accessed September 22, 2012.

3. If you ask an astronomer or astrophysicist about the color of the sun, you're likely to get as a response something like M. N. Perrin's "Calibrating the Color Temperature Relation: The B-V Color of the Sun," *Annales de Physique* 6 (1–2): 115–120 (1981), which isn't as clear as the CIE coordinates in describing what people see.

4. The *Munsell Color Laboratory Resources Page* has a particularly useful collection of Chromaticity functions and tables, at http://www.cis.rit.edu/mcsl/online/cie.php, accessed September 22, 2012.

5. On the ancient perception of the sun's color, I don't know of any one good source—the information is spread around books and websites on ancient art. One reference is Faber Birren, *The Story of Color: from Ancient Mysticism to Modern Science* (1941).

{ 22 }

Once in a Blue Moon

The phrase "Once in a Blue Moon" means "Rarely."[1] My understanding was that, on rare occasions, atmospheric conditions conspired to create a moon that really did look blue, but that these occurred very rarely, giving rise to the idiom. Some people held that a "Blue Moon" was a month that had two full moons in it—uncommon, but surely not as rare as a moon that really appeared to be blue. Then a few years ago an article came out which said that this interpretation was incorrect, and that a Blue Moon was something else altogether.

All of that is, broadly, true. But it's missing all the details and all the liveliness. And, critically for us, it skimps on the relevant optics. Here is the full story.

It's true that some people do call a month with two full moons a "blue moon." The idea has a recent feel to it, as if it's trying to come up with a situation in which the name "blue moon" could be applied to a rare event. But why would anyone want to invent such a convoluted explanation? Who cares if there is such a thing as a month one could point to and call a "Blue Moon"? The popular phrase could as easily exist without it.

As an article by Donald W. Olson, Richard Tresch Fienberg, and Roger W. Sinnott observes,[2] James Hugh Pruett, writing in the March 1946 issue of *Sky and Telescope*,[3] misinterpreted the work of an earlier *S&T* author and of the Maine *Farmer's Almanac* and noted that seven times in 19 years there are 13 moons in a full year, 11 with one full moon and one with two, This month with the unusual double moon he took to be the Blue Moon. The idea was repeated in later issues and received wider notice[4] when it was reported on the January 31, 1980, edition of the radio show *StarDate*. From there the usage spread. So the notion *is*, in fact a recent one, effectively dating from just over a quarter century ago.[5]

But, as those articles note, the real definition used by the Maine *Farmer's Almanac* is rather different. The true definition is that the "Blue moon" is the *third* moon that occurs when there are *four* moons in a season of the year.

Why? Why is there such a complex definition for what would seem to be a capricious name? The truth lies in realizing that this is not *a* Blue Moon, but *the* Blue Moon. In the almanac, which you can purchase every year, each full moon is given a name—*Hunter's Moon, Harvest Moon, Sturgeon Moon*, and so on. There are traditions of naming the moons all over the world, but the names used in the Maine *Farmer's Almanac* have a distinctly American ring to them. They include names unique to North America. Exactly how far back these go isn't clear, but they date to at least 1819.[6]

If there were an exact division of the days of the month into the days of the year—if, that is, the lunar year and the solar year coincided at beginning and end—that would be the end of the story. But they don't, of course. There's no fundamental reason for the length of a trip of the moon around the earth to be directly related to the time it takes for the earth to go around the sun (however those periods are defined). A lunar year has about 354 days, while a solar year has about 365.25 days. If one does not take some sort of measure, such as adding extra days, the months of the lunar year end up circulating through the solar year. This is precisely what happens with the Muslim calendar. The month of Ramadan can occur in any season. One easy alternative to constantly adding days is to occasionally add an extra month to bring the lunar and solar calendars back in synch. In the Hebrew calendar, which is a lunisolar calendar, an extra month is added at intervals to ensure that the correct months end up at the right seasons of the solar year. The extra month is called *Adar I*, and it occurs before the normal month of Adar, which is called *Adar II* on these occasions.[7]

The Maine Yankee Almanac does the same thing, adding an extra month whenever four months occur in the same season. Of course, all they are doing is changing the name of a full moon that is occurring. But instead of calling it "Hunter's Moon II" or anything like that, the extra month is called "Blue Moon," regardless of the season of the year when it occurs. In this way, year to year, Harvest moon always shows up in the fall, during the harvest season.

It's a remarkable system, considering how calendar reform is often a messy procedure, requiring the force of government or religion behind it in order to break people out of their well-trodden ruts to embrace a change. This system of Yankee months didn't require the force of the Roman Republic, or the Vatican, or the Muslim or Hebrew elders behind it. Exactly who is responsible for a system that works so well isn't clear. Perhaps it was the editor at the *Farmer's Almanac*. Perhaps the reason it was accepted so easily was that there were no great issues at stake. Nobody calculates salaries or payments or banking matters by lunar months in the United States, after all. And the *Farmer's Almanac*, being a useful and generally accepted guide, could easily proclaim its own calendaric system and get it bloodlessly enacted.

As for the phrase "once in a Blue Moon," the general opinion expressed now seems to be that it originally meant "never." A 1528 pamphlet *Rede Me and Be Not Wrothe* is the first citation found thus far, and the quote, modernized, reads "If we say the moon is blue, we must believe that it is true."[8] The *Oxford Dictionary of Idioms* ([1995] 2004, p. 21) notes that "the color blue was an arbitrary choice in this phrase. The Moon is Blue is recorded in the 16th century as a way of saying that something could not be true." The *Oxford English Dictionary* notes usages of "blue moon" as meaning something rare from 1821, 1876, and 1869.[9] John S. Farmer's 1890 book *Slang and its Analogues, Past and Present* gives examples from 1860, 1876, and 1884,[10] the last two being in the form "once in a blue moon." It observes that there are no known examples of the phrase prior to 1876, which it finds curious, and says that "A Blue moon, like the Greek Kalends, is something which does not exist."

And so, some see the progression from meaning "never" to "rarely" to the "correct" almanac rule to the modern "incorrect" two full moons in a month rule. Yet the fact that real blue moons *do* appear must give us pause. Surely this is evidence that "blue" was not an arbitrary choice, and that the very rare occasions of its appearance strongly suggests that "once in a blue moon" did *not* mean "never," but just what people have most often taken it for—a very rare occurrence.

In *The Road and the Inn* James John Hissey[11] writes:

> The moist atmosphere of the Fens appears to be productive at times of strange phenomena, for I was told that the moon, at certain times, has been observed to rise a pure blue over the Fens, with an effect most weird, as one may imagine; and that will-o-the-wisps are occasionally to be seen dancing and wandering erratically close above them. Curiously enough I had only just written this, when, on glancing at the *Observer* of January 23, 1916, my eye caught the following paragraph confirming the tale of the moon:
>
> "Seeing that the occasion had been so long waited for, the "blue moon" reported from East Norfolk seems to have been strangely unaccompanied by momentous events.... Though generally used as synonymous with the Greek Kalends, the Blue Moon is a recognized phenomenon. Brewer records one as having occurred on December 1883."
>
> It appears to me that this uncommon phenomenon may very possibly have originated the proverbial saying, "Once in a blue moon," meaning only very rarely, I take it. To account for this phenomenon is not easy.

In one sense, blue moons are *not* so rare. M. Minnaert, in his 1967 lecture[12] for the Optical Society that added meteorological observations to those already published in his book *The Nature of Light and Color in the Open Air*, noted that "Blue moons" are sometimes reported when the full moon is surrounded by brilliant orange or purple clouds a sunset. The moon's normal white seems blue by comparison.

But there is another sense in which the moon and the sun can appear very blue even without such contrast. And that, at last, is the true optical subject of this paper. A famous case of blue suns and moons occurred in September 1950 in Canada and Northern Europe, and another more recently in New Mexico on April 19, 1991. An excellent color photograph of the blue sun in the latter case appeared in the journal *Atmospheric Environment* in 1994.[13] What causes such blue suns and moons?

Normally, atmospheric scattering is dominated by gas molecules and random particles that follow Rayleigh's rule, in which the scattering varies with the inverse fourth power of the wavelength. This gives us our conventional blue sky, since short wavelength blue is preferentially scattered, and the sun appears red at sunset, since red is the least scattered of colors, especially through the longer atmospheric path at sunset. Something significantly different must happen to invert this order to scattering to cause blue to be the *least* scattered of colors.

Rayleigh scattering assumes widely dispersed noninteracting scatterers much smaller than the wavelength of light. The natural place to look for variations from this is to assume that these conditions are violated. Arguably the simplest way is to assume that one still has widely dispersed noninteracting scattering centers, but that they are no longer much smaller than the wavelength of light. For the simplest case, one assumes these to be spherical. The solution for scattering from spherical particles is, of course, Mie theory.[14] The problem with Mie theory is that the individual terms converge with glacial slowness. This problem is no longer an issue with the rapid recent increase in computing power, as given by Moore's Law, but it was a major problem even after computers had been in use for decades. It is worth noting that as recently as 1979 papers were still being presented on how to overcome this slow convergence by clever mathematical techniques.[15] Calculations of Mie scattering were difficult and complex, but there were at least 20 publications of such calculations by 1950.[16]

The general shape of the extinction versus the parameter

$$2\pi a|n-1|/\lambda$$

(where **a** is the particle diameter and **n** is the refractive index) looks very similar for most values, an oscillatory function that starts at zero, then goes through a series of maxima and minima, tending toward a nonzero limit As the index gets larger than 1 there are more complex ripples superimposed on this, and they reach closer to the origin, but the overall shape is the same.[17] As a general rule, the color of transmitted light through a considerable length of such scatterers is red if you are in a region of this plot where the slope is positive, and blue if you are where the slope is negative. At maxima the transmitted light is magenta, at minima it is greenish.[18]

One way to see the effect of Mie scattering, without being lost in the calculations and the endless convergents, is to observe, as G. F. Lothian did[19] that interference between rays passing through the diameters of such particles and through the air will experience maximum (minimum) opacity when **N** is odd (even) in the equation

$$2a(n-1) = \frac{N\lambda}{2}$$

This can be re-arranged to give

$$\frac{2\pi a}{\lambda}(n-1)=\frac{N\pi}{2}$$

The simple model works relatively well for the first "oscillation" or two, but gets "smeared out" over several periods. The extinction curve follows the same pattern as the extinction cross-section, which tends toward a limiting value of two.[20, 21]

References

Observations of Blue Moons and/or Suns

"The Blue Sun of 1950 September" R. Wilson *Monthly Notices of the Royal Astronomical Society* 111 pp. 478–489 (1951) and at http://adsabs.harvard.edu/full/1951MNRAS.111..478W, accessed September 22,

Bull, G.A., "Blue Sun and Blue Moon," *Met. Mag.* 80: 1–4 (1951).

Horvath, R., G. Mitzig, O. Preining, and R. F. Pueschel, "Observation of a Blue Sun over New Mexico, USA on 19 April 1991," *Atmospheric Environment* 28 (4): 621–630 (1994).

The Mid-Land Naturalist 7 (1884), p. 23, talks about a blue and a blue-green moon, and associates it with iron particles in the atmosphere, from meteoric or volcanic action. Earlier in the journal, it noted spectacular sunsets associated with the eruption of Krakatoa. http://books.google.com/books?id=KD84AAAAMAAJ&pg=PA23&dq=%22Blue+moon%22&hl=en&sa=X&ei=gE-8T5SWJ6-e6QGtw6RT&ved=0CE0Q6AEwAzge#v=onepage&q=%22Blue%20moon%22&f=false, accessed May 22, 2012.

Blue Moon/Sun Theory

Hogg, H. S., "Blue Sun," *J. Royal Astron. Soc. Canada* 44 (6): 241–245 (1950).

Lothian, G. F., "Blue Sun and Moon," *Nature* 168 (4286): 1086–1087 (December 22, 1951), Letters section, commenting on Paul and Jones letter.

Paul, Wiliam, and R.V. Jones, "Blue Sun and Moon," *Nature* 168 (4274): 554 (September 29, 1951), "Letters" section.

Pendorf, Rudolf, "On the Phenomenon of the Colored Sun, Especially the 'Blue Sun' of September 1950," *AFCRC Technical Report 53-7, Geophysical Research Papers #20* (April 1953) Geophysics Research Directorate, Air Force Cambridge Research Center, Cambridge MA. Online at http://www.dtic.mil/cgi-bin/GetTRDoc?AD=AD0007493, accessed September 22, 2012.

Pesic, P., "A Simple Explanation of Blue Suns and Blue Moons," *Eur. J. Phys.* 29: N31 (2008).

Porch, W. M., "Blue Moons and Large Fires," *App. Optics* 28 (10): 1778–1784 (1989).

Porch, William M., David S. Ensor, Robert J. Charlson, and Joel Heintzenberg, "Blue Moon: Is This a Property of Background Aerosol?" *App. Optics* 12 (1): 34–36 (January 1973).

Shute, C. C. D., "The Blue Moon Phenomenon," *Weather* 31 (9): 292–296 (September 1976).

Notes

1. Although many citations, especially from the late nineteenth century, suggest that it means "Never."

2. Donald W. Olson, Richard Tresch Fienberg, and Roger W. Sinnott, "What's a Blue Moon," originally in the pages of the magazine *Sky and Telescope* 97 (5):36–38 (May 1999), and now online at http://skyandtelescope.com/observing/objects/moon/article_127_1.asp or https://digital.library.txstate.edu/bitstream/handle/10877/4033/fulltext.pdf, accessed October 8, 2012.

3. James Hugh Pruett, *Sky and Telescope* 5 (53): 3 (March 1946).

4. As folklorist Philip Hiscock noted in another *Sky and Telescope* article, also entitled "Once in a Blue Moon," 97 (3): 52 (March 1999), now online at http://skyandtelescope.com/observing/objects/moon/article_377_1.asp, accessed October 8, 2012.

5. It is interesting, nonetheless, to note a relevant coincidence—in the March 31, 1888, issue of the journal *Notes and Queries: A Medium of Intercommunication between Literary Men, General Readers, Etc.*, "Daniel" of Plymouth asks "'Once in a Blue Moon'—What is the origin of this expression? What is a 'blue moon'?" to which the magazine replies, "The question remains practically unanswered." But the next query—placed there no doubt because it's also about the moon—is Ap. E. Coatham's question, "Moon Lore—Is there any folk-lore relating to the kind of winter which follows the occurrence of two full moons in the same month, as was the case in October last?" There was no reply (7: 248, no. 118, Saturday, March 31, 1888, Seventh Series, Vol. 5). Over 50 years before Pruett's error. 6.These names for the various months seem as if they might be some ancient pagan designations, but they're not—you won't find references to them before the nineteenth century. This hasn't stopped Pagans, Wiccans, and other "earth religion" groups from adopting and using them, however. Any good search engine will turn up plenty of lists like this: http://www.angelfire.com/de2/newconcepts/wicca/moons2.html, accessed May 22, 2012.

7. http://en.wikipedia.org/wiki/Hebrew_calendar, accessed September 22, 2012.

8. http://en.wikipedia.org/wiki/Blue_moon, accessed September 22, 2012, is the most accessible citation.

9. *Oxford English Dictionary*, 2d ed., Vol. 2, p. 325.

10. John S. Farmer, *Slang and its Analogues, Past and Present*, 260. Online at http://books.google.com/books?id=S7FZAAAAMAAJ&pg=PA260&dq=%22Blue+moon%22&hl=en&sa=X&ei=oES8T5G4N-Lo6AGI8PUn&ved=0CE4Q6AEwBDgK#v=onepage&q=%22Blue%20moon%22&f=false, accessed September 22, 2012, which attributes it to William Ernest Henley. The saying about the Greek Kalends is a subtle witticism—the *calends* was a Roman system, not used by the Greeks. "Until the Greek Kalends" became proverbial for something that would never happen. I confess that I had to look this up. General classical learning has declined since the nineteenth century.

11. James John Hissey, *The Road and the Inn* (1917), 206–207. Online at http://books.google.com/books?id=zNYxAQAAIAAJ&pg=PA207&dq=%22Blue+moon%22&hl=en&sa=X&ei=lUu8T9qvH4eY6QGe-dxP&ved=0CEEQ6AEwATgU#v=onepage&q=%22Blue%20moon%22&f=false, accessed May 22, 2012.

12. M. Minnaert, "Unusual or Neglected Optical Phenomena in the Landscape," lecture reprinted in *J. Optical Soc. America* 58 (3): 297–303 (1968).

13. H. Horvath, G. Metzig, O. Preining, and R. F. Pueschel, "Observation of a Blue Sun over New Mexico, USA, on 19 April 1991," *Atmospheric Environment* 28 (4): 621–630 (1994).

14. Although attributed to Gustav Mie and Peter DeBye, the theory of scattering from spherical particles of arbitrary size really has several claimants. See Nelson A. Logan, "Early History of the Mie Solution," *JOSA* 52: 342–343 (1962). For Mie Theory, see, for example, M. Born and E. Wolf, *Principles of Optics* (1959), chapter 13.

15. H. M. Nussenzveig, "Complex Angular Momentum Theory of the Rainbow and the Glory," *JOSA* 69 (8): 1068–1079 (August 1979) and S. D. Mobbs, "Theory of the Rainbow," *JOSA* 69 (8): 1089–1092.

16. H. C. van de Hulst lists all of the publications he was aware of up until 1957 in his book *Light Scattering by Small Particles* ([1957] 1981), 167–171, one of the standard references on the topic. Curiously, he does not list his own works, or that of A. N. Lowan, P. M. Morse, H. Feschbach, and M. Lax, *Scattering and Radiation from Circular Cylinders and Spheres* (1946) on those pages.

17. For some reason, many writers plot the extinction as a function of $\mathbf{2\pi a/\lambda}$ for a given refractive index, but fail to mention that they have done this. The factor of $|\mathbf{n-1}|$ is important to properly scale the result, however.

18. See Victor K. La Mer and Milton Kerker, "Light Scattered by Particles," *Scientific American* 188 (2): 69–77 (February 1953).

19. G. F. Lothian, "Blue Sun and Moon," *Nature* 168 (4286): 1086–1087 (December 22, 1951).

20. The effective cross-section is twice the geometric cross-section. See Van de Hulst, *Light Scattering by Small Particles*, 13–14.

21. More recent articles on diffractive arguments for the "Blue Sun" and "Blue Moon" are Peter Pesic, "A Simple Explanation of Blue Suns and Moons," *European Journal of Physics* 29 (3): N31 (2008); and Oleg V. Angelsky, Halina V. Bogatyryova, and Peter V. Polyanskii, "Interference Coloring of Regularly Scattered White Light," *Optica Applicata* 38 (2): 431–444 (2008), which references the original publication of the article this chapter is based on.

{ 23 }

Chromatic Dispersions

One of the more interesting optical effects is the *Christiansen Effect*, discovered by the euphoniously named Christian Christiansen of Denmark in 1884.[1] He found that if he placed glass ground into small fragments into a liquid having the same refractive index, but a different dispersion (that is, a different **dn/dλ**, so that the index changed at a different rate for the liquid than the solid), then the combination of the two would transmit light at the wavelength where the indices coincide, but would scatter at other wavelengths. The effect is rather dramatic, with the colorless liquid and solid collaborating to create a brilliantly colored composite having a single spectral hue. Christiansen used glass in a mixture of benzene and carbon disulfide, but many other combinations have been discovered, including solid-in-solid mixtures.[2]

Christiansen effect filters generally take advantage of the considerably larger **dn/dλ** often found in liquids—the larger the difference in dispersion, the narrower the bandwidth. One can make a very serviceable Christiansen filter using methyl salicylate (still available as a liniment at specialty drug stores), which has a very high dispersion, mixed with ethanol (one can obtain any refractive index between that of ethanol and methyl salicylate by varying the proportions of the two), mixed with most common glasses (microscope slides are very useful and readily available). For liquids, the change in refractive index with increasing temperature is typically 10 to 100 times what it is for solids, so you can "tune" the filter by heating or cooling it, changing the color of the passband. You can also tailor a Christiansen filter by obtaining samples of specialty glass from a supplier like Corning or Ohara and using index-matching fluid from a firm such as Cargill Specialty Liquids.

The basic relationships have been derived several times.[3] Although some workers claim that the scattering is due to multiple passes through grains causing phase shifts to accumulate, it seems very clear to me from studying grains of glass surrounded by index-matching fluid that the entire colored region occurs in regions where you have close to grazing incidence. This is particularly clear when you look at glass microspheres in fluid—the central region is quite clear and there is no observable deviation of the rays, or coloring. The index match is extremely good at most angles. It's only when you approach grazing incidence that you encounter a situation at which all wavelengths encounter an index mismatch that causes scattering. Unless the indices match perfectly, near grazing incidence all wavelengths but that one will scatter, and you have a ring of transmitted light corresponding to the wavelength at which the indices coincide. A little thought will convince you that there will inevitably be some surface of a grain of *any* shape for which you will have such near-grazing incidence. This is why no great care is needed to make a Christiansen filter—no matter what shape you grind your particles into, there will be enough grazing-incidence surfaces to guarantee satisfactory discrimination between index-matched and non-index-matched rays.

The use of Christiansen filters as passband filters was, I have been told, common before the Second World War, back when it was not easy to produce controlled layers of thin-film deposition.

If you look at a broad-area Christiansen filter made in a parallel-sided container, then no matter how well matched the indices are, in almost all cases a cell filled with a mixture of solid-phase and liquid-phase materials will only be effectively transparent for an object held right against it. The farther removed an object is, the more translucent the filter becomes. Over any significant distance, the filter is very poor at transmitting coherent images. It works very well as a method of removing unwanted wavelengths by scattering, and the recommended method of using such filters is to place them in a system where a broad collimated beam is passed through the cell. This fact, coupled with the problem of keeping a liquid-tight seal (especial with changing volumes) and the temperature sensitivity of the passband conspire to make Christiansen filters far less desirable than thin film stacks,[4] but before such thin-film filters became easy to manufacture, Christiansen filters were far easier to make.

My experience with Christiansen constructions has been that very small differences can cause enormous scattering changes. Packages of glass microspheres immersed in index-matching fluid invariably reveal index variations. I have had the same experience with plastic pellets (intended for injection-molding input feeds) immersed in index-matching fluid. In small quantities, or in round bottles they look like excellent cases of index matching, with the individual beads losing their identity to the fluid (plastics closely resemble liquids in their optical properties, and it's possible to find combinations with very nearly identical dispersion, so there is no obvious coloring), but placed in bulk in a flat-sided cell it is immediately clear that there is more than one phase present.

Variations in refractive index due to conditions of production or to stress-induced birefringence play a part in this imperfect index-matching. So does scattering due to surface contamination, fractures, and the like. Rayleigh had a way to minimize this problem—he ground his glass with an iron mortar and pestle, then boiled the resulting glass fragments in hydrochloric acid to dissolve any iron contamination, then dried and annealed the glass.[5]

Even with such treatment, however, it's hard to avoid the problems that can plague Christiansen filters. There is one method, however, that avoids the problems of both surface contamination, fractures, stress birefringence, and index variation from piece to piece. It's the only method I've personally been able to use to produce filters which are truly transparent.

Liquid-liquid Christiansen filters are made of two immiscible liquids which have the same index at one wavelength, but different dispersions. The individual particles of the phase consist of spherical droplets of one phase in the other. In fact, one will have droplets of each phase within the other. Because it's unlikely that two liquids suited to this purpose will have the same density, one will in general sit atop the other, and there will be an effective dispersion only near the interface unless the phases are vigorously mixed in some fashion, or else if the lighter phase is continually pumped to the bottom of the heavier phase. Such liquid-liquid filters have been called *Chromatic Dispersions*, which I like better than *Liquid-liquid Christiansen Filters*. There is the possibility of confusion with the use of the term to denote separation of light in an optical system because of the variation of index with wavelength, but I think that the context should make the correct meaning clear.

Despite these difficulties, the dispersions thus created allow transparent viewing. A simple chromatic dispersion can be made from turpentine and glycerine. At room temperature the passband is blue, but changing the temperature with a water bath (one should be careful, since turpentine is flammable) allows one to shift it to green.[6]

The scattering that comes when the phases are mismatched in refractive index has led to the suggestion to use Christiansen filters as power limiters in optical systems where one wishes to protect components (cameras, detectors, easily damaged optical materials, and the human eye itself, which is all of the above) from high intensity light. If the beam is focused into a Christiansen filter, then heating of the liquid, or nonlinear optical changes in index (which will happen much more rapidly) will change the index of one phase, and the resulting scattering will prevent the beam from continuing through the system.[7]

Christiansen filters show up in other contexts, as well. One would think that it's not possible to make a Christiansen filter in which air is one of the components, since air has a refractive index barely more than 1.0, and virtually all solid materials have indices greater than 1.2. In the infrared, the refractive indices tend to be much larger for transmissive materials than in the visible. In the vicinity of absorption bands the refractive index suddenly plunges in value, and so for many materials there are regions in the infrared in which the index actually drops to one. Accompanied, one must add, by a large absorption, so the material does not completely disappear. Nevertheless, light will pass through a thin layer of powdered magnesia, for instance (MgO_2) without scattering loss at about 12 microns.[8]

Notes

1. C. Christiansen, "Untersuchungen über die optischen Eigenschaften von fein verteilten Körpern," *Ann. Phys. Chem.* 23, 24: 298–306, 439–446 (1884, 1885).

2. Some Christiansen filter "recipes" are given in W. G. Driscoll and W. Vaughan (eds.), *Handbook of Optics* (1978), 8-103–8-106; L. Auerbach, "Some Simple Christiansen Filters," *Am. J. Phys.* 25 (7): 440–442 (1957) and L. Auerbach, "More Christiansen Filters," *Am. J. Phys.* 28 (8): 743 (1960).

3. C. V. Raman, "The Theory of the Christiansen Experiment," *Proc. Indian Acad. Sci.* A29: 381ff. (1949); C. V. Raman and K. S. Viswanathan, "A Generalized Theory of the Christiansen Experiment," *Proc. Indian Acad. Sci.* A41: 55ff. (1955); R. H. Clarke, "A Theory of the Christiansen Filter," *App. Opt. 7*: 861ff. (1968). Clarke's paper contains an interesting typo that has not yet been corrected: "What have been labeled equations **(A4)** and **(A5)** are actually a single equation that has been run onto two lines."

4. Methods of constructing useful Christiansen filter arrangements, including temperature stabilization, appear in many places, including John Strong, *Procedures in Experimental Physics* (1938), 373; E. D. McAlister, "The Christiansen Light Filter: Its Advantages and Limitations," *Smithsonian Misc. Coll* 33 (7): 00-00 (1935); and U. Wojak et al., "Christiansen Filters for the Far Ultraviolet: An Old Spectral Device in a New Light," *Appl. Opt* 26: 4788ff. (1987),

5. Lord Rayleigh, "On an Improved Apparatus for Christiansen's Experiment," *Phil. Mag.* 35: 373ff. (1918). See also Auerbach, "Some Simple Christiansen Filters."

6. Other chromatic dispersion include hydrofluorosilicic acid plus potassium chloride, and partially saponified ethyl cinnamate, both cited by R. W. Wood, *Physical Optics*, 3d ed. (1934), 115–117. A recent interesting case is tetrahydrofuran plus aqueous sodium thiosulfate solution with hydropropylcellulose added. See Kento Okoshi et al., "The Christiansen Effect of Brightly Colored Colloidal Dispersion with an Amphiphilic Polymer," *J. Colloid and Interface Sci.* 263: 473–477 (2003).

7. See, for instance, George L. Fischer, Robert W. Boyd, Thomas R. Moore, and J. E. Sipe, "Nonlinear-optical Christiansen Filter as an Optical Power Limiter" *Opt. Lett* 21 (20): 1643–1645 (1996).

8. John Strong, *Procedures in Experimental Physics* (1938), 375; Walter G. Driscoll (ed.), *Handbook of Optics* (1978), 8–106; R. B. Barnes and L. G. Bonner, "The Christiansen Filter Effect in the Infrared," *Phys. Rev* 49: 772 (1936); Vereshchagin and Borisevich, "A Study of Infrared Dispersion Filters" (1968).

{ 24 }

The Eye in the Spiral

> The knobs he had taken for eyes were his first concern. A close examination of their surfaces revealed nothing, so he carefully tried to detach one from its stem. It finally cracked raggedly away, and proved, as he had expected, to be hollow.... As a sudden thought struck him, Cunningham held the front part of the delicate black bit of shell in front of his eyes; and sure enough, when he looked in the direction of the spaceship, a spark of light showed through an almost microscopic hole. The sphere WAS an eye, constructed on the pinhole principle—quite an adequate design on a world furnished with such an overwhelming luminary.

These lines come from the September 1945 issue of *Astounding Science Fiction*, from the story *Uncommon Sense* by Harry Clement Stubbs, writing under his usual pseudonym "Hal Clement." Clement was famous for his scientifically rigorous stories, often involving alien creatures living in very different environments than the ones humans inhabit, producing surprising differences. In this case, the apparent pinhole eyes turn out to have olfactory organs, rather than a retina. For creatures living in a high vacuum (as these did), the very long mean free path of emitted molecules allows such a pinhole "nose" to create an image of the source in an odd but workable amalgam of sight and scent.[1]

But even a real pinhole eye can only exist in a science-fiction environment, surely? Evolutionary biologists have postulated (and modeled) a developmental sequence for vision starting with light-sensitive patches, which enclose to become pinhole systems, but these are soon displaced by vision systems using lenses, which can concentrate more light and produce better and more detailed images. By this route, vertebrates and cephalopods independently evolved eyes with the same structure—cornea, iris, lens, and retina within a spherical capsule.[2] Surely the inefficient pinhole eye, with its very low sensitivity or, if the hole is enlarged to admit more light, poor imaging quality must give way to the modern camera-type eye, which combines both good light-gathering ability and high-quality imaging.

But the pinhole eye lives, in three different cases. One of these is the "pits" of Pit Vipers (*Crotalinae*) and of some Pythons (*Boidae*). Their pit organs are deep indentations with a somewhat restricted entrance (whose size is much larger than one would like to call a "pinhole") and a floor lined with infrared-sensitive cells. Because of the large aperture, these pit organs can "see" rather blurry images of thermal sources. This is very useful to a creature that hunts for warm-blooded prey, especially when combined with ordinary light-sensitive vision. It's easy to understand why snakes might have these crude pinhole eyes—they give another sense in addition to standard vision, in a different wavelength regime.[3, 4]

The second is *Tridacna Maxima*, the Giant Clam, which has a pinhole eye with a 90 micron diameter aperture on a 500 micron diameter eye. Experimentally, based on animal response, and from the optics of the situation, *Tridacna* has a resolution of about 20°. Michael F. Land, who studied the eye at length, determined that it allowed the clam more time to react than mere shadow sensitivity alone (just light-sensitive patches, with no pinhole eye structure), allowing the clam time to retract its sensitive (and vulnerable) mantle. On the other hand, its relative insensitivity acts as a behavioral filter, keeping the clam from constantly opening and closing its shell, as a more developed eye would have the clam do. Its pinhole eye seems like a "Goldilocks" solution to its environment—sensitive enough to allow it to protect itself when danger looms, but not so good that it responds to every stimulus.[5]

But the third case is a surprise, in view of the often-lauded similarity between the octopus eye and the human eye. A different cephalopod, a distant relative of the octopus, has a simple, lensless pinhole eye. It's the most unusual of the cephalopods, the one with the spiral shell that inspired submarine christeners and exercise machines—the Chambered Nautilus.

The Nautiloids developed as a separate group during the Cambrian period and spread throughout the world, but were replaced by modern-day coleoid cephalopods, except for the Chambered Nautilus. Today its range is restricted to the Southwestern Pacific (although there is evidence that it may have ranged as far as the Indian Ocean in classical times). It is a nocturnal creature, but has an active and unusual optical sense.[6]

Its eye consists of an elongated oval, about 10 mm by 15 mm, with the long dimension horizontal. It measures about 9 mm from pupil to retina, and the pupil is an ellipse measuring on average about 2 mm high by 1 mm wide (although, like the pupils of other animals, it can vary in size, but keeps approximately the same relative dimensions). There is neither lens nor cornea, and the pupil is simply an opening to the outside, so that the interior of the eye is filled with seawater. There is a trough running vertically downward from the pupil, which is filled with cilia that can propel water, and it's been speculated that these hairs are used to provide a current of water across the pupil that keeps it free of debris.

W. R. A. Muntz and U. Raj in 1983 made models of Nautilus eyes with different pupil sizes and used them to photograph standard eye charts.[7] With the smallest pupils (1 mm high × 0.4 mm wide), horizontal features could be clearly discerned, but vertical features were very nearly blurred out by the larger pupil dimension. For larger pupil sizes (2 × 1 and 2.8 × 1.7) structural details were very nearly lost along both directions, but the locations of objects were nevertheless very evident.

The Nautiluses certainly do use their vision to orient themselves, both horizontally and with regard to their surroundings. Muntz and Raj placed Nautiluses in a tank surrounded by drums having regularly spaced stripes. When they rotated the drums, the Nautilus rotated to follow the stripes. At any rate, they did when the stripes subtended angles of 22.5° and 11.25°. When the stripes subtended angles of 5.5° and 2°, the Nautiluses were unable to follow them. This accords well with the visual acuity of about 5° that Muntz and Raj calculate based on the eye model.

One would expect that the retina of the Nautilus, like that of Tridacna, would have a coarse system of sensing cells to match its coarse optical resolution, but the density of receptors is actually comparable to that of other cephalopods that can resolve objects with 17 minutes of separation.

The visual system of the Nautilus is something of a mystery. It has coarse pinhole resolution better for horizontal than vertical objects, yet has a densely packed retina. It lives in a low-light environment, yet can apparently compete with other creatures having more highly developed eyes, with better light-gathering capability (Muntz and Raj estimate that comparable eyes, with lenses, would gather two orders of magnitude more light). Vision is clearly important to it—besides the above features, it shows the ability to learn based upon vision[8]—yet its optical system seems extremely primitive. Early researchers, finding no lens in their specimens, were concerned that they might have lost them in the process of preserving and preparing their specimens.

What is the secret of Nautilus vision? Do they hunt for bioluminescent prey? (Muntz and Raj raise the possibility, but point out objections as well. Later work, however, tends to support this.[8]) Can their brains process the information from the blurred images passing over the dense receptors and deconvolve to obtain high-resolution information?[9] Are the features remnants of an earlier complexity that has somehow resisted genetic drift? Or is there some simpler but nonobvious explanation for the anomalous Nautilus eye?[10]

Notes

1. Hal Clement (Harry Clement Stubbs), "Uncommon Sense," *Astounding Science Fiction*, September 1945. Reprinted in *Space Lash (Small Changes)* (1969) and *The Best of Hal Clement* (1979).

2. See, for instance, M. F. Land and R. D. Fernald, "The Evolution of Eyes," *Annual Review of Neuroscience* 15: 1–29 (March 1992).

3. Eric A. Newman and Peter H. Hartline, "The Infrared Vision of Snakes," *Scientific American* 246 (3): 116–127 (March 1982).

4. George S. Bakken and Aaron R. Krochmal, "The Imaging Properties and Sensitivity of the Facial Pits of Pit vipers as Determined by Optical and Heat-Transfer Analysis," *J. Experimental Biology* 210: 2801–2810 (2007).

5. Michael F. Land, "The Spatial Resolution of the Pinhole Eyes of Giant Clams (Tridacna Maxima)," *Proc. Royal Soc. London B* 270: 185–188 (November 28, 2002).

6. Peter Douglas Ward, *In Search of Nautilus* (1988), 21–23.

7. W. R. A. Muntz and U. Raj, "On the Visual System of *Nautilus Pompilius*," *J. Exper. Biology* 109: 253–263 (1984). See also W. R. A. Muntz, "Spectral Sensitivity of *Nautilus Pompilius*," *J. Exper. Biology* 126: 513–517 (1986).

8. W. R. A. Muntz, "Effects of Light on the Efficacy of Traps for *Nautilus Pompilius*," *Marine Behaviour and Physiology* 24 (3): 189–193 (1994).

9. Peter Douglas Ward, *The Natural History of the Nautilus* (1987), 72–73.

10. Ward, *The Natural History of the Nautilus*, 7.

{ 25 }

Retroreflectors

Retroreflectors surround me. I have reflecting facets built into my car taillights and attached to my bicycle. I have flexible retroreflecting strips sewn onto my backpack and my pant-cuff ties for cycling. The many traffic signs along my streets are retroreflecting, and there are retroreflecting "cat's eyes" built into my highways. Many of the white highway lines are coated with reflecting paint. People mark boundaries of their property with reflectors and identify potential obstacles with them. Retroreflectors make highway barriers and street signs extremely visible in the dark, with no light source or power source of their own. They've been around all my life.

And yet, retroreflectors are a surprisingly recent invention. As recently as two centuries ago the very concept of retroreflection had still not been fixed. This is a little surprising, since there are examples of retroreflectors in nature—animal eyes (especially cat's eyes) seem to shine in the dark with reflected light. And the phenomenon of *heiligenschein*, wherein one's shadow perceived against a background of freshly dewed grass seems to have a glow about the head, are both natural examples of retroreflectors—cases where an object will reflect light back along very nearly its incident path, regardless of where the light is coming from. And yet, in these two cases, it seems as if people didn't really realize until recently that they were cases of reflection at all.

That might seem incredible, but in *The Philosophical Transactions of the Royal Society of London* for 1799 there is reported a series of investigations[1]:

> the experiments which brought the subject of nervous structure under consideration, were made on the eye, and were in some measure connected with the observations made in former lectures; they were instituted with a view to ascertain the cause of the luminous appearance frequently observed in the cat's eye.
>
> The illumination so conspicuous in the eye of the cat, and of many other animals, when seen in an obscure light, has attracted the attention of every common observer. Philosophers also have paid particular attention to it, and have endeavored to investigate the cause. On this subject there have

> been two opinions: one, that the illumination arises from the external light collected in the eye, and reflected; the other, that there is a quantity of light generated in the organ itself. Professor Bohn, at Leipsick, made experiments which proved, that when the external light is wholly excluded, none can be seen in the cat's eye. These experiments were favourable to the first opinion; but the brightness of the illumination is so great, that it appeared to exceed any effect which could be produced through the medium of the retina; so that some other source of light was thought necessary to account for the phenomenon: this circumstance gave support to the 2nd opinion.

This reads like some enormous and straight-faced joke, some characteristically dry piece of British humor, but it is soberly recorded in the archives of the Royal Society of London, the most learned and august scientific organization of its day. On the one hand, it is gratifying to know that they went about investigating the subject very carefully and thoroughly, using the true Scientific Method. But one can't help being amazed that the best scientists of their day were considering whether or not cat's eyes were glowing from their own internal light or not, even after having been presented with experimental evidence that this was *not* the case. The mind boggles.[2]

Of course, it's not as if even the next round of experiments, as described in the report, settled the matter completely and for everyone. Belief in the innate luminosity of animal eyes survived into the twentieth century. As Bergen Evans noted in his book *The Natural History of Nonsense*:

> The eyes of many animals are popularly assumed to emit light. This may be an echo of the Greek "emanation hypothesis" of vision—the ancient belief that seeing was accomplished by the sending out from the eye of slender threads, thinner than gossamers, that touched the object seen. But it is more likely an erroneous assumption that the light reflected from the eyes of animals otherwise invisible in the dark is not a reflection but an emanation.[3]

The other natural occurrence of retroreflection occurs when spherical droplets of dew are placed on the grass and observed in the light of the rising sun (or, more rarely, the setting sun), with the observer's shadow falling on the field of grass. For reasons we shall give below, such an arrangement will produce an intense reflection very close to the direction the light coming from, and so the observer will see the shadow of his or her head surrounded by a blaze of light. This has been called the *Heiligenschein*, German for "halo," although it more literally means "holy shining." There are several phenomena that have been called by this name, but since the middle of the nineteenth century this term has been used specifically to refer to retroreflection due to light focused by nearly spherical particles onto a background, then redirected by the same particles back toward the source. I have diligently looked for this effect in my own dew-covered grass, but have never seen it there—my grass is evidently of the wrong sort, and the dew drops adhere strongly to the sides of the blades, flattening out into elongated half-ellipsoids, and give no retroreflection effect. For more on this, see below. I have, however, seen this effect from glass beads deliberately sprinkled, or from glass beads broken free from reflective paint, or from the glass bead-type reflective paint itself.

This and other optical effects, including the *opposition effect* and the *glory*, probably gave rise to the artistic conventions of the nimbus, the aureole, and the halo, where the heads of particularly holy individuals are surrounded by a bright light. The convention appears in Christian art, Buddhist art, and Australian art, among others. The optical effects referred to above produce a bright light (in the case of the glory, a rainbow-like colored light) around the shadow of the head of the observer. Notably, it doesn't appear around the shadows of the heads of one's companions, since these effects are the results of strong retroreflection, so the observer appears to him- or herself to be more favored than one's companions. But before you get too swollen a head, consider that, if you take a picture of the effect, the Holy Light will be centered on your blessed camera, not on you.[4]

I know of no ancient references to these effects, but there is a fascinating testament to it in the autobiography of the artist Benvenuto Cellini (1500–1571). At the end of the first volume of his autobiography, he remarks[5]

> ever since the time of my strange vision until now an aureole of glory (marvellous to relate) has rested on my head. This is visible to every sort of men to whom I have chosen to point it out; but those have been very few. This halo can be observed above my shadow in the morning from the rising of the sun for about two hours, and far better when the grass is drenched with dew. It is also visible at evening about sunset.

This report would probably have been written off as another of the hallucinations or visions Cellini reports in his book, had it not been for that clinching detail of the brightness being greater when the grass is covered with dew. As a result, this is regarded as the first report of *heiligenschein*, and the effect has (rarely) been called the Halo of Cellini.[6]

Some observations on the silvery light reflected from the leaves of colewort plants (better known in America as *collard*, the cabbage relative that is the source of collard greens) were made by Thomas Melvill, one of the first to observe spectral lines (see Chapter 10, "First Light—Thomas Melvill and the Beginnings of Spectroscopy"). He observed that the reflecting water droplets appeared to be suspended above the surface of the leaves, and that if the drops were to "wet" the leaf, the silvery appearance retreated.[7]

An attempt at explaining the effect was made in 1825 by Heinrich Wilhelm Brandes, in which he accounted for it using both external and internal reflections. His explanation has persisted through the years, being reproduced in the authoritative 1910 *Meteorologische Optik* of J. M. Pernter and in the 1976 edition of the *Encyclopedia Britannica*.[8]

The first reference to the optical effect by the name *heiligenschein* that I have found dates from 1834, echoing Cellini[9]:

> In the 61st number of the Vienna "Jahrbuch der Literatur" there is a review of a work, containing an account of a newly-observed phenomenon, a luminous border surrounding shadows under certain circumstances. The work is by Dr. C. Garthe, and is entitled "Uber der Heiligenschein," literally "Upon the Saint's-Shine." A newly compounded word in German (as we suppose) and a title which seems to us somewhat affected. It refers, we presume, to the halo represented in paintings around the heads of saints.
>
> The facts which the author professes to have discovered are these:
>
> If a person stand with his back to the sun in the morning or towards sunset, so that his shadow may fall upon a field covered with dew-drops, he may see a luminous border round his shadow, which if his head be uncovered will be most distinct round the shadow of the head. The breadth of this border varies according to the circumstances from five inches to six feet.
>
> The phenomenon can be observed only during three hours after the rising, or three before the setting of the sun.

The first correct explanation of the phenomenon seems to be due to E. Lommel[10] in 1874, who felt that the surface of the grass must be within a focal length of the droplet, and assumed that hairs on the surface of the grass blades kept it there, instead of wetting the grass. Lommel's theory met with the criticism that the retroreflected light ought to be green, like the grass, instead of white, and this may have caused some doubt about the explanation.[11] R. A. R. Tricker, in the late 1960s, performed experiments to show that the retroreflected light is very unsaturated and appears nearly white with any color of background behind it, even black.

J. O. Mattson and C. Cavallin[12] observed that not all the grasses on which *heiligenschein* was observed even *had* hairs. They suggested that the drops retained their spherical shape and simply rested on the surface. Alistair B. Fraser of the Department of Meteorology at Pennsylvania State University appears to be the first to try to produce a quantitative theory of the effect and its geometry.[13] Fraser observed a similar, but not identical, effect on the leaves of trees, chiefly conifers, that he termed "sylvanshine," and theorized (matched with his observations) that you can get a *heiligenschein* effect even if your drop was in contact with the surface of the leaf, and even if it partly wetted it, as long as the contact angle was not too large. With partial wetting, the appearance changes, hence his *Sylvanshine* appears different from *heiligenschein*. And, presumably, the grasses on my lawn have far too small a contact angle. Fraser's comments on trees exhibiting the effect are worth quoting:

> A nonexclusive list of conifers that exhibit the sylvanshine contains the blue spruce, hemlock, juniper, . . . cedar . . . the Douglas tree . . . the white fir, and (to my delight) the Fraser fir.

For most of the nineteenth century, the *heiligenschein* effect was little more than a curiosity. But with the advent of the automobile, there was a new and practical use for the effect.

The use of lines or raised stones to mark roads has apparently been used for thousands of years, in ancient Rome and in seventeenth-century Mexico. Putting a stripe on the road in the United States is credited to Edward Hines, a road commissioner in Wayne County Michigan in the first years of the twentieth century. The practice evidently helped keep order and reduce accidents, and soon spread through the country. At first, some places used black lines, but these weren't visible at night. White paint was better, but still not as visible as one would like at night, even with automobile head lamps. There were reportedly some experiments with the new radium paint, which glowed without external power supplies, but this would have been an expensive solution.[14] Finally, in the 1930s an article appeared in the journal *Canadian Engineer* entitled "Luminous Marking for Highways," which suggested the use of glass beads atop white paint for high visibility. The journal *Engineering News-Record* in 1942 observed that some 500 miles of roads had been painted with glass bead paint in the Philadelphia area. The cost was greater (Virginia DOT estimates three times greater), but the paint with glass beads mixed in was much more abrasion resistant. And, of course, it was demonstrably more visible. *Popular Science* for January of 1942 reported that "marking of Highway center stripes with a light-reflecting paint is a new step forward making the roads safer at night."[15]

Why should it take so long for retroreflectors to become a practical item? Evidently it was the growing use of the automobile. With many people traveling great distances at night over possibly unfamiliar roads, some means of more effectively and visibly marking roads were needed. Attempts to use simply white paint, or even luminous paint were not effective, and clearly street lighting alone was insufficient. The new automobiles had directed headlights, with parabolic reflectors, and this provided their own source of light. The field was ripe for retroreflectors, which had never really been needed before, and wouldn't have been practical. Both the need and the technology became available at the same time. There had been horse-drawn carriages with lanterns before this, but they moved more slowly, and the lanterns threw out less light. Neither the need nor the light technology had been present.

Within a decade reflective paint with mixed-in glass beads were being advertised and widely used to paint centerlines and boundaries on roadways. In these cases the paint was formulated so that the glass beads stay partially submerged in the paint. In some cases, the beads sat atop the paint (frequently, in fact, the beads were scattered after a layer of paint had been laid down and rested atop the paint). This was a poor situation, and the beads in these circumstances can easily be torn out of their setting. Eventually the paint loses its reflective ability. The ideal was for the beads to be at least half buried in the paint. In these cases, the combination retroreflected light, but was robustly held in place.

Pretty clearly, we have here a situation where the bead is not suspended above the surface, but is actually in direct contact with it. The setup is different from even Fraser's *Sylvanshine*, because the glass beads remain nearly spherical, not "wetting" their background. What is certainly happening here is that the light passes through the glass surface and is refracted toward the back surface, but instead of being refracted again upon leaving the glass sphere, the light hits a white layer of paint at that surface, and is scattered in all directions. Any light that is scattered into the same cone that it was in when it struck the paint will be retroreflected back along the original path. Moreover, light that is nearly within that cone will be approximately retroreflected—which is actually what is desired. No one, after all, wants the light to go back to the headlight of the car. It only serves a useful purpose if it is not precisely retroreflected, but is a little away from the angle of retroreflection, so that it gets to the eye of the driver.

This glass bead retroreflection is an easy and economical means of making a retroreflector, in other words, because it is such an incredibly tolerant optical construction. Unlike the optics used in precision lens combinations and optical instruments, the separation between the glass bead (or the water droplet) and its background can be any distance between right against the glass bead and a bit more than one focal length away. The retroreflector will function in any position in that range. Moreover, it works very well for "in direct contact," which is easy to engineer, by simply putting the bead in contact with the paint. And finally, the bead need not even be perfectly spherical.

The same principle finds use outside of retroreflection for safety purposes. As R. A. R. Tricker observed, it has long been used in art and craft work. "There are several cases of the back reflection of light by transparent spheres, or other solid or liquid bodies with curved surfaces," he wrote, in his *Introduction to Meteorological Optics*, "They include what used to be known as 'pastinella work'—one of the bastard arts in which objects like cushions and curtains were decorated by painting on them with oil colours and then sprinkling small transparent spheres over them while the paint was still wet" (p. 24).

I have searched a long time for any other references to "pastinella work," without any success. Even with the awesome capabilities of various Internet search engines and the various tools available from sites such as Google, I can uncover no references which are not due to Tricker, or myself in referencing him. I'm forced to conclude that the use of the term is either restricted in region or time so much that it has not left a footprint on the Internet, or else he was mistaken in his use of the term. His observation that this is used in craft work is, however, quite valid. Craft stores sell tiny glass beads. Even the goddess of household crafts, Martha Stewart, has a signature line of them for sale. I find reference to the practice in *The Crafty Witch: 101 Ideas for Every Occasion* by the intriguingly named Willow Poison, in *Embellished Fashions* by Mickey Baskett, and in *Terry O's Easy Embellishing: Instantly Accessorize Everyday Objects* by Suzanne Chase and Terry

O. I have seen beads (for the amateur beader) that have been embellished with such glass microbeads, giving them a luminous appearance.

But the use of retroreflecting beads is not limited to the hobbyist and the "bastard arts." Since 1964 artist Mary Ann Corse has been using glass microbeads on her monochromatic artworks, which are now highly sought-after and respected. Ms. Corse is part of a California-based art movement called the Southern California Light and Space movement. "But my paintings are not reflective!" she insisted.[16] "They create a prism that brings the surface into view. I like that because it brings the viewer into the light as well."

As a physicist, I might quibble with her use of the term "prism," but there is no doubt that the interaction of the light with the microbeads and the background is complex, as the above history and analysis has shown, and has resisted a proper understanding for quite a while. Mary Corse, like any good artist, has explored the features and effects of her medium, and used it to good advantage. "From one angle, each painting looks perfectly opaque, blank, and flat. Move a few steps closer or away, and silvery gray vertical bands appear like invisible ink exposed to ultraviolet light. Move again and the bands start to shimmer and take on depth, dissolving back into white a few steps later. No matter how you move, you feel the glow." Photographs of her works completely fail to capture the experience, although photographs taken at different locations and looking along different directions will show you how the appearance can vary as you move around. But they won't show the ever-changing appearance or the bright glow. As *Times* writer Linda Yablonsky notes, "There's no passive looking to be done here."

But these reflective glass beads are only one type of retroreflector. There are many different types of device that will send light counterpropagating toward its source, and they have different properties and capabilities. The heiligenschein/glass bead retroreflector has the virtue of being easy to fabricate and forgiving of errors in its construction, but suffers from vulnerability—even optimally "buried" glass beads get pulled loose in time through wear—and imprecision in the direction of retroreflection. It sends the light back in a relatively broad angle range. Other designs can do better in this regard.

The French Colonel of Artillery and later University Professor of Geodesy and Topography (at Metz and Fountainbleau) Charles-Moyse Goulier is credited with the invention of the Penta Prism. This is a block of glass with two faces set at an angle of 45° to each other. Next to each of these are two faces, usually square, that meet at a right angle to each other and each of them makes a angle of 112.5° with the other two faces. The prism can work as it is, with only four faces, but the 45° angle is far from the rest of the prism and gives it an unwieldy long "tail," and all that extra glass adds weight without contributing anything optically, so it's usually cut off, adding a fifth face, hence "penta" prism. It's symmetrical about that cut-off face, but is not a regular pentagon having equal sides and angles. Light enters through one of the square faces, bounces off first one, then the other of the long faces with the 45° angle between them,[17] then exits through the other square face, so that the exiting light makes a precise right angle (90°) with the incoming light. This may not seem very impressive—I can, after all, bend light through a right angle like that using only a single mirror. Or, if I wanted to use a Total Internally Reflecting prism, a 45° right angle prism will do it. But if my mirror or right-angle prism isn't positioned at just the correct angle, the reflected light will be deviated by some angle other than 90°. The genius of Goulier's Penta Prism is that, no matter what angle the prism is placed at (within limits—the light still has to go in through one of those square faces), the outgoing beam *always* makes an angle of 90° with the incoming light.

Goulier announced his discovery in 1864. He used it in range-finders for measuring distance, a useful device for a colonel of artillery. Its ability to keep that constant 90° angle, regardless of angle or handling, would definitely appeal to a soldier used to working in the chaotic circumstances of war.[18] I'm told that his prism is called a "Goulier Prism" in his honor, but I've only ever heard or read of it as a "penta prism."

The same principle of reflection from a prism at a constant angle between incoming and outgoing beams was used somewhat later by A. Beck in 1887. Instead of an angle of 90° between the incoming and outgoing beams, he obtained an angle of 180°—that is, the light was reflected backwards along the direction it came. He did this by using an angle of 90° between the two faces of his prism (exactly half of the angular difference between the incoming and outgoing beams, just as Goulier's prism used an angle of 45° between the faces, which is half of the 90° deflection) But Beck didn't simply use a right-angle prism to reflect his light in a plane. He found that if he made a three-dimensional prism, equivalent to slicing off one corner of a glass cube and reflecting the light from all three faces, that the beam will be retroreflected back along its original path, no matter how the prism is rotated in space.[19]

Because of the way it is made, this item is commonly called a "Corner Cube" Some people call it a "Cube Corner." Professor Rudolf Kingslake of Rochester insisted that this makes much more logical sense than "Corner Cube," a term which he regarded as an abomination. He consistently used "Cube Corner" in publications from the 1920s until his death in 2003. This prism is almost invariably called by one of these two names, or simply called a Retroreflector. By rights, it ought to be a Beck Prism, but that name is extremely rarely used (as it is, for instance, in US Patent No. 4,032,236). Beck himself called it a *TripelSpiegel*, or "Triple Mirror," and specified that, when the surfaces met at right angles, it was a *Zentralspiegel*.[20]

The Corner Cube sends its light back much more nearly in the opposite direction than the glass bead reflector, and concentrates the light in a narrower beam, but it has much tighter tolerances and is much more difficult to make. Its only natural counterpart is the salt crystal, which can make perfect cubes, although I have never heard of anyone using these as retroreflectors, or of reporting the effect in them.

The problem, of course, is that corner cubes are technically much more demanding than glass spheres to make. They must be laboriously cut and polished. There is some leeway in the tolerance on the angles at the edges, since we don't really need perfect retroreflection, and would actually prefer somewhat imperfect reflection, but it still requires a good surface finish. Corner Cubes are still made and marketed as high precision instruments by several companies. They can also be made out of three separate mirrors that meet at right angles.[21] It was possible to cast or press glass into rows of corner cube shapes, but the real breakthrough for this technology was when it became possible to mass-produce optical plastics. Since the outline shape of a corner cube is an equilateral triangle, and you can use equilateral triangles to tesselate a plane, if you cover a sheet with such corner cubes you can create a retroreflecting sheet with no dead spaces, giving you a completely efficient retroreflector. You can also cut off the corners of those triangles and make hexagons, which can also tessellate the plane. Today a great many retroreflectors are made of hard plastics and soft, covering bicycle reflectors, automobile tail-lights, flexible strips on joggers' outfits, fabric ties around cyclist's legs, on children's Halloween costumes, firefighter's jackets, and thousands of other uses.

There is a third major type of retroreflector. Imagine the spherical glass type of retroreflector, in which the glass beads brings the light toward a focus. When the beads are adhered directly onto a surface, as with the beads mixed into road paint, the light never does come to focus, since the focal point of a sphere made of water, plastic, or most glasses falls outside the sphere.[22] It would be much better, in place of using a perfect sphere, to choose the front radius of curvature so that, for the material being used, the light came to focus at the rear surface. Furthermore, the location of the focal point as you go off axis does not fall on a flat plane at the focal point location. (A fact photographers are all too aware of, and they pay a lot of money for complex lenses engineered to keep the light in focus across the surface of their image plane.) The focal surface curves, forming the *Petzval surface*, named after the nineteenth-century physicist and optical pioneer Joseph Petzval, who analyzed it and designed a camera lens to overcome the effect. If you place a curved surface with the Petzval radius at the back of the device, and use its Fresnel reflection or (even better) silver it to improve the reflection, you will have a device that takes incoming parallel light and retroreflects it back along the direction it came, and do so much more efficiently than scattering glass beads on a painted surface. You will, in fact, essentially reproduce the structure of a *real* cat's eye, which brings light to focus on a curved surface and retroreflects it back through the lens.

Why should a cat's eye do this? What purpose is served by making the cat's eye an efficient retroreflector? In fact, nature didn't set out to make the eye a retroreflector at all. The goal of the eye is efficient vision, and the round eyeball with its corneal lens in front is a simple and efficient design for all vertebrate eyes (and quite a few invertebrates, such as the cephalopods). Just as with the *heiligenschein*, reflection from the retina will generally be gathered and re-collimated by the corneal lens of the eye. Of course, the goal of the retina is not to reflect light efficiently, but to absorb it, so that the eye's owner can see. So the retina isn't ideally designed to be a retroreflector. And human eyes aren't. What light is retroreflected is reddish, and contributes to the "red eye" effect often seen in amateur photographs of friends.

Since some light is coming back, the human eye obviously isn't completely efficient at absorbing the light. Neither are the eyes of other animals. In order to increase the efficiency, some animals—cats, most notably, and the wolves Bergen Evans wrote about, have a reflective layer called the *tapetum*, which lies *behind* the retina and reflects un-absorbed light back through the retinal cells, giving them a second chance to absorb the light.[23] Of course, throughout the evolution of cats there weren't intense sources of light around, or photographers ready to grab images. The bright retroreflection is a happy (or unhappy) but "unintended" consequence of the evolution of the eye for efficient use of photons. It's just our good luck that this can now result in hundreds of photographs showing up on the LOLcat pages on the Internet.

A recent study comparing the retroreflecting intensities of actual animal eyes (they used cow's eyes and deer eyes) against glass bead and paint retroreflectors and plastic tessellated corner cubes found that, while high-quality tessellated corner cubes could be up to six times as efficient as animal eyes, the animal eyes were more efficient than the glass bead retroreflectors or low-quality tessellated plastic corner cubes, and the animal eye retroreflection was collimated into a very narrow 1° full width at half maximum.[24, 25]

Artificial "cat's eyes" made of glass for human road safety seem to have been independently invented at least twice. Frederick Walter Madeley Lee worked for General Motors in Luton, Great Britain, and applied for a patent for "Improvements to and relating to road signs and the like" on January 20, 1932, patent application 1728/32. He constructed prototypes, using both clear and colored glass, suggesting that different colors might be used to designate different hazards or situations. Unfortunately, he could not afford to continue funding his application efforts and had to abandon the effort.[26]

The generally acknowledged inventor of the device is Percy Shaw of Halifax, Yorkshire, a road mender. Shaw was unaware of Lee's efforts, and came up with the invention, according to him, when driving home on a foggy night from Bradford to the Boothtown area of Halifax. At one stretch of the road there was a drop-off, and one could normally tell where it lay by looking for the reflection from railway tracks. On this night in 1933, however, the tracks had been removed for repairs. Shaw was warned about the location of the edge of the road by reflections from the eye of a cat crouching there, which stimulated his thinking about the design possibilities.[27] Apparently he told a less melodramatic version, in which he was inspired by seeing reflective road markers, and wanted to make better ones. Shaw's design (unlike Lee's, which consisted of a glass bead attached to a mount, to be screwed into a wooden surface), consisted of an iron holder for four glass beads set in rubber, two pointing in each direction. The iron mount gave a tough setting that would protect the beads from damage, and was meant to be mounted in the road. His patents (Nos. 436,290 and 457,536) were registered in 1934, and he set up Reflecting Roadstudds Ltd. to manufacture them.

If, instead of changing the radii of curvature of the front and back surfaces to make the light come to focus at the rear of the sphere, we varied the index of refraction throughout the sphere, we could make a lens in which all incoming parallel light is brought to focus at the rear surface, then reflected backwards through the same optical part. This was what Rudolph K. Lüneberg did in the 1940s. He showed that a dielectric sphere with a radially-varying index of refraction following the formula

$$n(\rho) = \sqrt{2 - \left(\frac{\rho}{r}\right)^2}$$

n(ρ) = the refractive index as a function of **ρ**
ρ = the radial distance from the center of the sphere
r = the radius of the sphere.

would have the properties described above.[28] In his honor, such a lens is called a *Luneberg Lens* (dropping the *umlaut* over the *u*, since most American typesets don't use them). Such a lens is physically unrealistic, since the index varies from an extremely high value of **2** at the center to a value of **1** at the outside (and, hence, there would be no refraction upon entering or leaving the sphere). No simple solid material has a refractive index of 1 in the visible (some fluoride glasses and compounds have indices less than **1.3**), and it is difficult to get radial refractive indices to vary as neatly as one would like, or to go as high as **2**. The Luneberg Lens would seem to be doomed to be only a theoretical curiosity, impossible to construct physically. But that is not the case.

Luneberg lenses are easy to approximate in the range of radio waves and microwaves, where concentric layers of varying effective index are substituted for a continuously changing gradient, and where one can approach an index of unity. Such radar retroreflectors are valued over simple corner cubes because the cross-section is unvarying with angle for a Luneberg lens.

In the visible, even if the index cannot be made vanishingly small, reasonable approximations can be made using only a few steps of index, with a ball of one refractive index buried within two concentric shells of other material, and with the radii and thicknesses chosen to compensate for the refraction at the outer surface of the device (references). As with longer-wavelength Luneberg lenses, optical Luneberg lenses offer a uniform cross-section.

Although it is difficult to make a reliable three-dimensional smooth gradient, two-dimensional Luneberg lenses have been made and used for planar coupling. Recently, metamaterials have been used to make Luneberg lenses.[29-37]

These are probably the most common types of retroreflectors in use, and probably constitute most of the commercial systems in use. But these are by no means the only retroreflectors.

A glass or plastic sphere can act as a retroreflector without any material behind it, and without altering either the front or back radius of curvature, provided its refractive index lies between the square root of two and two. As R. A. R. Tricker very elegantly shows,[38] this will result in light at a limited range of radii away from the center of the sphere being perfectly retroreflected, but not all of the rays that enter the sphere. As a result, when you look at such a sphere with the light behind you, it will only seem to "glow" in a ring around the center, and will be dark in the center, and outside the ring. Water has a refractive index of **1.33**, too low for this effect. Most glasses have indices somewhat higher, but common glasses at about **1.5** will only light up the outer rim of the sphere. You need high index glass, or a spherical flask full of a high index material like Carbon disulphide (n = **1.63**) to really see the effect. Such high index materials tend to be expensive, and aren't really necessary, so this one finds no practical use.

One may also use a parabolic reflector. Light coming in parallel to the optical axis (the axis of symmetry) of the parabola will be reflected through the focal point, then will strike the opposite side of the parabola and be redirected parallel to the axis again. Unfortunately, this, too, will leave a dark center, and it only really works for light very nearly parallel to the parabola's axis, unlike the three major types, which will accept light from a very broad range of angles and retroreflect them.

The *Handbook of Optical Systems: Survey of Optical Instruments*, Volume 4[39] contains an entire chapter devoted to retroreflector designs, using such methods as an axicon (a conical surface, acting as a reflector or, in this case, a refractive element) with a reflecting flat behind it. A reflecting axicon with an apex angle of 90° would also acts as a retroreflector.

Retroreflectors have been used as laser end mirrors with varying degrees of success. It was a common practice to use them as mirrors on ruby lasers in the early 1960s. More recently, a cat's eye reflector has been used as laser end mirrors in a standard cavity, and shows remarkable stability.[40]

Notes

1. "The Croonian Lecture. Being Experiments and Observations on the Structure of Nerves" Ev.[Everard] Home, F.R.S. An. 1799 Vol. LXXXIX p. 1 *The Philosophical Transactions of the Royal Society, from their commencement in 1655 to the year 1800*, abridged Vol. 18 (from 1796 to 1800) (1809) reprinting Vol. 89, pp. 1–12. Online at http://books.google.com/books?id=Ey5WAAAAYAAJ&pg=PA431&dq=%22cat's+eye%22&hl=en&sa=X&ei=S4CeT_GKIIe36QGN5PSADw&ved=0CHIQ6AEwBg#v=onepage&q=%22cat's%20eye%22&f=false, accessed April 30, 2012. The original *Philosophical Transactions* article is here: http://www.jstor.org/stable/10.2307/107021, accessed April 30, 2012.

2. Dr. Everard Home was still writing on this subject almost 30 years later. In his *Lectures on Comparative Anatomy* (1828), Vol. 3 of 4, pp. 243–244, he starts out with precisely the

same wording, and goes on to note that the experiments were "attended with difficulty; for when the apartment is darkened, and nothing but the light from the cat's eye seen, the animal, by change of posture, may immediately deprive the observer of all light from that source. This was found to be the case whether the cat, the tiger, or hyena was the subject of the experiment." One is simultaneously disappointed by the doctor—still wondering if cat's eyes glowed of their own light—and awed by the experimentalist willing not only to test this hypothesis, but to sit in the dark with a tiger or hyena to prove it. Accessed here: http://books.google.com/books?id=XHlD8pNzBxkC&pg=PA243&dq=%22cat's+eye%22&hl=en&sa=X&ei=2IGeT9GUObGd6AH3-7WiDw&ved=0CFMQ6AEwAQ#v=onepage&q=%22cat's%20eye%22&f=false, accessed April 30, 2012.

3. Bergen Evans, *The Natural History of Nonsense* (1946), 67.

4. See, for example, Plate II.1.B on page 26 of R. A. R. Tricker's excellent *Introduction to Meteorological Optics* (1970).

5. *The Autobiography of Benvenuto Cellini*, trans. J. Addington Symonds (1910), 262.

6. The persistence of the effect when the grass *isn't* covered in dew is likely due to the *opposition effect*, as Bruce Hanke points out in *Theory of Reflectance and Emittance Spectroscopy* (2012), 220–227.

7. Thomas Melvill, "On the Silver-like Appearance of Drops of Water on the Leaves of Colewort," reprinted *J. Royal Astron. Soc. Canada* 8 (4): 238–241 (July-August 1914). Online at http://articles.adsabs.harvard.edu//full/1914JRASC...8..231M/0000238.000.html, accessed October 27, 2012.

8. "Hof" in *Gehler's Physikalisches Wörterbuch* (1826), Vol. 5, pp. 439ff., cited, with comments on its longevity, by Alistair Fraser, "The Sylvanshine: Retroreflection from Dew-Covered Trees," *Applied Optics* 33 (21): 4539–4547 (July 20, 1994). See also Tricker, Meteorological Optics, 28 and 44, who disproves the assertion, but doesn't properly attribute it.

9. *Select Journal of Foreign Periodical Literature*, Vol. 1 for 1834, pp. 228 and 229.

10. "Ueber den Lichtschein un den Schatten des Kopfes," in *Annalen der Physik und Chemie* (1874), pp. 10–21, cited in Fraser 1994, note 8.

11. Tricker, *Meteorological Optics*, 33–35.

12. J. O. Mattson and C. Cavallin, "Retroreflection of Light from Drop-Covered Surfaces and an Image-Producing Device for Registration of This Light," *Oikos* 23: 285–294 (1972), cited in Fraser 1994, note 8.

13. See Fraser 1994, note 8.

14. Virginia Department of Transportation Manual on Pavement Marking 2012 http://www.virginiadot.org/business/resources/Materials/MCS_Study_Guides/bu-mat-PaveMarkCh2.pdf, accessed April 2012. See also *Traffic Engineering* 47 (1): 64 (1977).

15. Myron M. Stearns, "Mile for Mile, Your Chances of a Bad Accident are Three Times as Great After Dark as by Day," Popular Science 140 (1): 128–131 (January 1942). http://books.google.com/books?id=DCcDAAAAMBAJ&pg=PA127&dq=%22reflecting+paint%22&hl=en&sa=X&ei=-aOgT4S2EsL1gAeW9dyqCQ&ved=0CGAQ6AEwAzgK#v=onepage&q=%22reflecting%20paint%22&f=false, accessed May 1, 2012.

16. Linda Yablonsky, "Mary Corse," New York Times, February 24, 2012. Stored online at http://tmagazine.blogs.nytimes.com/2012/02/24/artifacts-mary-corse/, accessed May 3, 2012.

17. No silvering of the faces is needed, The incident light is completely reflected, due to the angle of incidence and the refractive index of the glass—an effect called *Total Internal Reflection*.

18. http://www.treccani.it/enciclopedia/charles-moyse-goulier/, accessed May 3, 2012; and *Metro Manual: A Handbook for Engineers Containing Technical Information Regarding the Construction, Adjustment and Use of Transits, Tachymeters, Theodolites, Alidades, Levels, Etc. Manufactured by Bausch and Lomb Optical Co.*, issued by Bausch and Lomb, Rochester NY (1915), 194–196 Available at http://books.google.com/books?id=djA7AAAAMAAJ&pg=PA196&dq=%22penta+prism%22&hl=en&sa=X&ei=Cc-eT63fGsn56QGStcTuDg&ved=0CGYQ6AEwAw#v=onepage&q=%22penta%20prism%22&f=false, accessed May 3, 2012.

19. Beck's article "Ueber einige neue Anwendungen ebener Spiegel" appeared in *Zeitschrift für Instrumentenkunde*, 380–389 (November 1887). Beck's work is also cited in Rudolf Straubel's 1906 US Patent No. 835,648 on behalf of the German optical firm, Carl Zeiss.

20. Robert B. Nielsen and Xiao Jing Lu, "Retroreflection Technology" in Proc. SPIE 5616 *Optics and Photonics for Counterterrorism and Crime Fighting* pages 47-60 (2004).

21. Sometimes called a "hollow cube," and useful for wavelengths that are absorbed by glasses. Note, also, that Goulier's "penta Prism" can be realized using two mirrors with a 22.5° angle between them, for the same reason.

22. In order to focus at the surface of the sphere, the index must be 2 for the paraxial case.

23. Most explanations of the reflections from cat's eyes mention the tapetum and its role in reflecting the light, but leave the reader confused—why should nature put a layer of reflecting cells at the back of the eye to reflect the light back out, when the eye really wants to *absorb* that light? They don't make clear the essential role of effectively increasing the absorption length of the retinal by placing the reflective surface *after* the light-absorbing cells.

24. Nathaniel R. Greene and Brian J. Filko, "Animal-Eyeball vs. Road-Sign Retroreflectors," *Ophthalmic and Physiological Optics* 30 (1): 76–84 (January 2010).

25. Incidentally, a report comparing glass-sphere retroreflectors to corner cubes found that corner cubes were superior for long-distance reflection, but that glass microspheres (not "cat's eyes") were better for nearby retroreflection (since they reflected more light away from strict counterpropagation). See George H. Seward and Pamela S. Cort, "Measurement and Characterization of Angular Reflectance for Cube-Corners and Microspheres," *Opt. Engineering* 38 (1): 164–169 (January 1999).

26. Frederick Lee's story, along with documentation and pictures of his prototypes, is available on a website maintained by his family. See http://www.catseyes.com/, accessed May 5, 2012.

27. Design Museum website for museum show on Percy Shaw's invention, November 2006. Located at http://designmuseum.org/design/percy-shaw, accessed April 2012.

28. R. K. Luneberg, *Mathematical Theory of Optics* (1944), 189–213.

29. Andrea DiFalco, Susanne C. Kehr, and Ulf Leonhardt, "The Luneberg Lens in Silicon Photonics," *Opt. Express 19* (6): 5156–5162 (2011).

30. F. Zernike, "Luneberg Lens for Optical Waveguides," *Opt. Comm. 12* (4): 379–381 (December 1979).

31. John Hunt, Talmage Tyler, Sulochana Dahr, Yu-Ju Tsai, Patrick Bowen, Stéphane Larouche, Nan M. Jokerst, and David R. Smith, "Planar, Flattened Lunenberg Lens at Infrared Wavelengths," *Opt. Express 20* (2): 1706–1713 (January 16, 2012).

32. V. D. Shargorodsky, V. P. Vasiliev, N. M. Soyuzova, V. B. Burmistrov, I. S. Gashkin, M. S. Belov, T. I. Khorosheva, and E. Nikolaev, "Experimental Spherical Retroreflector on Board the Meteor-3M Satellite," *Proceedings of the 12th International Workshop on Laser Ranging*, Materna Italy (2000).

33. V. D. Shargorodsky, V. P. Vasiliev, M. S. Belov, I. S. Gashkin, and N. N. Parkhomenko, "Spherical Glass Target Microsatellite," *XV International Laser Ranging Workshop* Canberra, Australia (2006).

34. John P. Oakley, "Whole-Angle Spherical Retroreflector Using Concentric Layers of Homogeneous Optical Media," *App. Optics 46* (7): 1026–1031 (March 1, 2007).

35. N. C. Anheier, B. E. Bernacki, N. A. Klymyshyn, K. Krishnaswami, and C. P. Rodriguez, "FY 2008 Miniature Spherical Retroreflectors—Final Report," *U.S. Department of Energy Report PNNL-18344* (February 2009).

36. Bruce E. Bernacki, Norman C. Anheier, Kamman Krishnaswami, Bret D. Cannon, and K. Brent Binkley, "Design and Fabrication of Efficient Miniature Retroreflectors for the Mid-InfraRed," *SPIE Proc. 6940—SPIE Defense and Security Conference 2008 Infrared Technology and Applications XXXIV*, Paper 6940-30.

37. Ettore Colombini, "Design of Thin-Film Lunenberg Lenses for Maximum Focal Length Control," *App. Optics 20* (20): 3589–3593 (October 15, 1981).

38. Tricker, *Meteorological Optics*, 36–41.

39. Herbert Gross, Fritz Blechinger, and Bertram Achtner, Handbook of Optical Systems: Survey of Optical Instruments, Volume 4 (2008).

40. Zhiguang Xu, Shulian Zhang, Yan Li, and Wenhua Du "Adjustment-free Cat's Eye Cavity He-Ne Laser and its Outstanding Stability," Opt. Express 13 (14): 5565–5573 (July 11, 2005).

{ 26 }

Yes, I was Right! It *Is* Obvious!

In his best-selling book *Chaos,* James Glieck expounds on the "tough guy" nomenclature employed by physicists and mathematicians. Problems that are truly important and original are "Deep." Everyone would love to discover and solve a Deep Problem. Problems that are straightforward and easy to solve are "Trivial." The problems that lie between these extremes are "Obvious." A reasonably competent scientist can work out an Obvious problem. But, in this sense, "Obvious" doesn't necessarily mean what most people think of as obvious.

There's a classroom anecdote that makes the point pretty clearly. It goes like this: One day, a professor is lecturing his graduate student class in some subject, and he writes on the board a series of equations, explaining how to get from one to the next. But he leaves out the step between the last two equations, saying, "It's obvious."

He starts to go on, but one student raises his hand and says that he might be slow, but the reasoning behind that step isn't obvious to him. The professor looks back at the equations, starts to speak, looks at it again, and ponders it. Suddenly he turns and leaves the classroom. The students talk in muted whispers. A minute goes by. Five minutes. Ten minutes. The professor returns, a smile on his face.

"I was right," he exclaims. "It *is* obvious!"

I had my own experience with the Obvious several years ago in grad school. I was doing work with the spectra of color centers—point defects in alkali halide crystals that often have absorption bands in the visible, rendering the otherwise colorless crystals blue, purple, or yellow. As a result, most papers on the topic feature graphs of absorption spectra. In many papers, the spectra are given in terms of wavelength. In others, the spectra are given in terms of wavenumber or electron volts. (For some reason, an awful lot of these were from the Soviet Union.) A few papers, trying to straddle both cultures tried to have it both ways. They'd have a wavelength scale on the bottom and a wavenumber scale on the top. But, of course, that still favors one side or the other. You can be linear in one of these, but not in the other. (No one has ever tried the experiment of not being linear in either). That usually makes one scale the preferred one, but you could always read off the value on the other scale, if you had to.

Exactly where the peak occurred wasn't a purely academic issue to me, however. I was calculating concentrations of defects. I was also looking at emission spectra and trying to calculate gain coefficients. The wavelength peak emission or peak absorption enter into these and other calculations, so I needed to know where they were. And I realized something interesting.

It made a difference how you were plotting the data. If you plotted it in wavelength space you got one answer, but if you plotted it in wavenumber space you got a different one. The shift wasn't large, and probably didn't affect any of my calculations in a meaningful way, but it was unmistakably there. And it was interesting.

I mentioned this thing I'd stumbled upon to one of the Professors, and said it might be worth a brief note or article in a journal. He didn't believe it, at first.

> "You mean to tell me," he said, "that if I had a spectrophotometer calibrated in wavelength and measured the peak location I'd get one value, but if I measured it on a machine calibrated in wavenumbers I'd get a different answer?"
>
> "Yes," I replied.
>
> "And if I converted them by $\boldsymbol{\sigma = 1/\lambda}$ I wouldn't get the same result?"
>
> "Nope."
>
> "Well, that doesn't make any sense."

We discussed it for a long while, and I rolled out my demonstration of the matter. It doesn't qualify as a proof, since it only looks at one example. Assume an absorption spectrum that is a Gaussian in wavelength space:

$$F(\lambda) = Ae^{-\left(\frac{\lambda-\lambda_0}{\Delta\lambda}\right)^2}$$

We can obtain the distribution in wavenumber space by using the rule that:

$$F(\sigma)d\sigma = F(\lambda)d\lambda$$

This is required if we want the areas between corresponding points to be equal.

$$F(\sigma) = F(\lambda)\frac{d\lambda}{d\sigma}$$

$$F(\sigma) = -\frac{1}{\sigma^2}F(\lambda)$$

It's reasonable, for purposes of comparison, to scale both of these to similar scales. Both distributions can be set to a dimensionless value of 1 at $\boldsymbol{\lambda = \lambda_o}$ and $\boldsymbol{\sigma = 1/\lambda_o}$

$$F(\lambda) = e^{-\left(\frac{\lambda - \lambda_0}{\Delta\lambda}\right)^2}$$

$$F(\sigma) = \frac{1}{\sigma^2 (\lambda_0)^2} \exp\left[-\left(\frac{\frac{1}{\sigma} - \lambda_0}{\Delta\lambda} \right)^2 \right]$$

Taking the derivative of the first of these and setting it equal to zero gives the location of the maximum at $\lambda = \lambda_0$, where the function has the value of 1. Taking the derivative of the second expression with respect to wavenumber and setting the result equal to zero produces the following expression for the maximum:

$$\sigma_{max} = \frac{-\lambda_0 + \sqrt{\lambda_0^2 + 4(\Delta\lambda)^2}}{2(\Delta\lambda)^2}$$

To first approximation, this is:

$$\sigma_{max} \cong \frac{1}{\lambda_0} \left[1 - \left(\frac{\Delta\lambda}{2\lambda_0} \right)^2 \right]$$

You can visualize the mathematics of the above by imagining the function of the absorption peak as a series of glass partitions set between glass walls. Between each pair of partitions there's water placed, with its height indicating the absorption function at that point. Let's say the original plotting is linear in wavelength space. We have partitions at, say 300 nm, 310 nm, 320 nm, etc., all equally spaced. Now we change the plot over to one in wavenumber space. We convert the wavelength of each marker into wavenumber using $\sigma = 1/\lambda$. Now we slide the partitions around so we have a scale that's linear in wavenumber. (We assume that you can slide these partitions around without losing the fluid inside.) The partitions were evenly spaced in wavelength space, but they won't be in wavenumber space—they'll be getting farther apart as you go to longer wavenumbers. Nevertheless, the integrated area between the partitions has to remain constant—that's why we represented that area with a fixed volume of water. Some pairs of partitions will get closer together, so the height of the water between them will increase. Others will get farther apart, so the height will decrease. In this way, the shape of the absorption spectrum in wavelength space changes into the absorption spectrum in wavenumber space. As you make the separations between the bins smaller and smaller you gradually approach a better and better depiction of the smooth and continuous function.

Visualizing the spectrum in this way, as a series of bins with preserved volumes, rather than as a point-by-point transformation, it is easy to see how the peak location not only *can* shift, but *must* shift, unless the spectrum is a delta function. If we let $\boldsymbol{\lambda_o}$ = **555 nm** and $\boldsymbol{\Delta\lambda}$ = **50nm**, then the maximum in wavenumber space is **0.00178741**, or **1/559.5 nm**, or **4.5 nm** from what it "should" be.

Once we'd gone through all this, I asked the professor if he thought it would make an interesting article.

> "No," he said, shaking his head, "I don't think so."
> I was stunned.
> "Why not?" I asked.
> "Well," he said, "It's just too obvious."

Oh, well, at least *this* time I got an article out of it.

This was the very first article I wrote as a Contributing Editor for *Optics and Photonics News*, helming the *Light Touch* column (I had written a *Light Touch* article almost a decade earlier on Retroreflectors, but I had no direct association with the magazine then. A greatly expanded and rewritten version appears as the previous chapter). I was somewhat nervous about how it would be received. I was surprised and daunted when the piece prompted a *huge* amount of correspondence. Many people wrote in to tell me that I was clearly and obviously mistaken, including one former president of the Optical Society of America. That was all distressing and made me question my stand and my math. But about as many wrote in to praise my perceptiveness and to assure me that they had been trying to push home the same point themselves. This latter group included *another* former OSA president. So it balanced out. The staff at OPN was delighted simply by the fact that the piece generated so much response. They took this as a sign that the article (and hence the magazine) was being read and talked about. It's the print media version of "any publicity is good publicity."

Nevertheless, I am apprehensive about negative criticism, not only because of my personal feelings, but because I want to achieve a correct understanding (and to direct my readers to such, as well). Thus the criticisms are worth reviewing.

One of the criticisms was that if one took the derivative of the integrated area, expressed both in wavelength space and wavenumber space, that it could be shown that the extremum in wavenumber space occurred at the same wavelength as the maximum. Starting with

$$\int_{\sigma_1}^{\sigma_2} F_\sigma d\sigma = \int_{\lambda_1}^{\lambda_2} F_\lambda d\lambda$$

He then took the derivative of each side, using $\mathbf{d/d\sigma = (d\lambda/d\sigma)d/d\lambda = -\lambda^2 d/d\lambda}$, and obtained

$$F_\sigma(\sigma_2) - F_\sigma(\sigma_1) = -\int_{\lambda_1}^{\lambda_2} \lambda^2 F_\lambda'(\lambda) d\lambda$$

So that the wavenumber distribution is

$$F_\sigma(\sigma) = -\int \lambda^2 F_\lambda' (\lambda) d\lambda$$

This will be equal to zero at $\boldsymbol{\lambda} = \mathbf{0}$ or at $\mathbf{F}'(\boldsymbol{\lambda}) = \mathbf{0}$. The first of these is trivial, but the latter is the maximum of the wavelength distribution. The maximum of the wavenumber distribution thus falls where the maximum of the wavelength distribution occurs, as one expects.

But there are problems with this argument. In the first place, the left side of the last expression isn't the derivative of the curve in wavenumber space (which would be $\mathbf{F}'(\boldsymbol{\sigma})$), but the value of the curve itself, so setting the right side equal to zero doesn't correspond to an extremum of the curve in wavenumber space. But this is trivial, because the real problem is that the derivative is improperly taken on the right side. One does not take the derivative of the expression inside the integral sign, but undoes that integral to leave the argument of the integral. The correct result is:

$$F_\sigma(\sigma_2) - F_\sigma(\sigma_1) = -\lambda^2 [F_\lambda(\lambda_2) - F_\lambda(\lambda_1)]$$

In order for this to be true for all values of $\boldsymbol{\lambda}$ and $\boldsymbol{\sigma}$ we must have

$$F_\sigma(\sigma_n) = -\lambda^2 F_\lambda(\lambda_n)$$

This is the result I'd obtained above.

Another criticism is much more subtle. It argues that I have made the assumption that the plot of both the curve in wavelength space and wavenumber space constituted a Distribution Function, the area under which is conserved. This, it was pointed out, is shown by my argument by analogy with the "containers of water" and is implicit in my development of the connection between the function in wavelength space and wavenumber space. But there is no reason to suppose that the curve represented really is such a function, the area under which must be the same regardless of the units it is measured in. If the function is (as in the example the professor made his argument about this) an absorption function, where the transmission on a scale of 0 to 100 percent is plotted either in wavelength or wavenumber space, then the area certainly is not conserved. In a real dual-beam spectrometer instrument the value is simply a ratio between two beams (a reference beam and a beam passing through a sample cell holding the material under test), and this ratio is not a Distribution Function. In fact, if you measure the absorption at a collection of wavelengths and plot them, then convert the wavelengths into wavenumbers and plot them, the maximum absorption clearly occurs at the wavenumber corresponding to the wavelength at which the maximum absorption occurs. And thus am I proven wrong.

It is true that the form the result is expressed in does make a difference, and I must confess to not being specific enough about this. If you are looking at an absorption spectrum on a scale of 0 to 100 percent, then all of this is true. But I was concerned mainly with two other types of spectra—emission spectra and absorption expressed as optical density, and I must confess that these are the types of display I had in mind. My work was on the absorption and emission of defect centers in crystals, and I was looking at the emission spectra, which gives the intensity of emission for a given excitation, and which is measured by placing a detector near the sample, or by imaging the emitted light onto the detector with a lens or mirror, or by using an integrating sphere with a detector in it. In any of these cases, the strength of the signal is a measure of the number of photons, and the number of photons *is* going to be the same, whether we measure the emission as a function of wavelength or wavenumber. There is, in other words, a connection between the function in wavelength and in wavenumber space, and the number of photons per unit wavelength or wavenumber space (which is the emission function divided by the energy per photon, given by $\mathbf{hc/\lambda = hc\sigma}$). So the connection really exists between the photon number emission function, rather simply the emission function itself. That is my error.

What about absorption? I took my measurements in optical density, rather than simple absorption, and so did most of the papers I was consulting. There is an advantage in doing this. In the first place, simple absorption tends to "flatten out" near 100 percent when the sample is optically dense, and you can't tell the difference between 99 percent and 99.9 percent absorption. And you can't tell what the shape of absorption is by looking at a plot. You can't see where the peak is. Plotting things on a logarithmic scale retains the characteristic shape of the absorption. The manufacturers of glass filters used to provide logarithmic plots of absorption vs. wavelength, plotted in black against axes plotted in red. Then they provided a red filter overlay with a scale plotted on it. By moving the scale upwards or downwards you could see what the internal optical density (the logarithm of the transmission, ignoring the reflection losses at the entrance and exit faces of the filter) was as you varied the thickness of the filter.

The optical density also provides a useful tool if you are creating and manipulating defect centers. Knowing the thickness of the sample and the optical density and the width of the peak (and some relevant physical quantities) you can tell how many defect centers are present. The area under the curve is tied to the number of absorbing centers present, and, again, this means that it is a Distribution Function, which must have a definite relationship between the absorption as expressed in wavelength space or wavenumber space. It was, in fact, my having to perform such calculations and converting between papers that expressed the result sometimes in wavelength space and sometimes in wavenumber space (and which would plot it with a scale linear in, say, wavelength on the bottom and nonlinearly in wavenumber on the top) that started me thinking about this entire issue.

In fact, although the argument about the spectrometer looks compelling on its own, most spectral lineshapes are not arbitrary functions with peaks appearing in eccentric places, but can be expressed pretty well in terms of mathematical functions that are tied to the broadening mechanisms, with many of them being Gaussians as I assumed in my original development (these are caused by Doppler broadening in gases, and by vibrations of the atoms in the crystal lattice in my solid-state case) or Voigt lineshapes, which are the convolution of a Gaussian with a Lorentzian. Once you can express and comfortably fit your spectral peak with a mathematical function, you certainly expect the function transformed into the inverse space to follow the same mathematical rules as that function, and the peak will shift.

So the situation, as originally expressed in my column was incorrect, because it didn't specify things clearly enough. The spectrum, if an absorption spectrum, ought to be measured in optical density. Ideally, the emission spectrum ought to be a photon number spectrum for conservation of area. But even if it is the emission intensity, the peaks in wavelength and wavenumber space will almost certainly not coincide unless it is effectively a delta function.

{ 27 }

Edible Lasers

When I was in Grad School, there were stories circulating about a "Jello Laser"—that someone had actually made a laser using Jello. Because, as everyone said, *anything* will lase if you "hit it hard enough." Searching through the *Science Citation Index* and *Physics Abstracts Index* (in those pre-Internet days, before computer databases), I located "Laser Action of Dyes in Gelatin" by T. W. Hänsch, M. Pernier, and A. L. Schawlow.[1] The paper revealed that it wasn't *really* Jello—they had simply put laser dye in Unflavored Gelatine—that non-dessert stuff they sold in odd-shaped boxes with a picture of a cow on the label, that was supposed to be good for your fingernails. But, the rumor mill went on, somebody *did* succeed in making a Gin and Tonic lase (it was the tonic water or the quinine that did it, they said).

I was grateful, then, when Theodor Hänsch published "Edible Lasers and Other Delights of the 1970s"[2] in 2005, setting the story straight. We learned that Arthur Schawlow was the source of the saying that "anything will lase if you hit it hard enough" and that he really *had* suggested trying to make colored, flavored gelatine lase. Unfortunately, none of the twelve flavors they bought and prepared actually lased. In the spirit later exemplified by *MythBusters*, he pointed out that Sodium Fluorescein was "almost non-toxic," which is how they ended up mixing it into the gelatine. The published note only detailed these later, successful experiments. The rumors about trying to make ordinary Jello lase, however, were apparently true. No word on the Gin and Tonic laser, though.

This got me thinking about whether it was really true that you could get anything to lase, and if there were any other Edible Lasers. It's been over 35 years since the letter by Hänsch et al., you'd think someone would have turned up something else. For the purpose of this essay, I assume "edible laser" to refer only to the lasing medium—I'm not going to require that anyone come up with a fully contained edible laser. And I'll allow that the medium might not be easily edible in the condition it's in for lasing. So it might have to be cooled, heated, ground up, or otherwise subjected to a temperature or size change—but no chemical changes, mixtures, or solutions. I think that's fair. Also, it really has to be eaten and at least partly digested—no grinding up inert glass or plastic laser hosts. And the actual lasing material ought to be edible. That disqualifies cases like the purported "X-Ray Laser" of Kepros, Eyring, and Cagle, in which the Copper Sulfate embedded in the unflavored gelatine was supposed to be doing the lasing.[3]

That doesn't leave many possible candidates. Most solid state lasers are out, since you can't digest most of the glass or crystal hosts. That leaves out almost all gas lasers, too—noble gases like Argon, Xenon, and Krypton freeze out at temperatures well below that of liquid nitrogen, Neon isn't much better, and metal vapors condense to, well, metals. The cadmium out of a HeCd laser isn't edible. So what does that leave?

A determined search through reference books, databases, and patent literature failed to turn up much. One place to start is with laser dyes. As Schawlow noted, Sodium Fluorescein isn't *very* toxic. Neither, for that matter, are Coumarin dyes. One expert I talked to noted that some dyes, like sulforhodamine dyes, exit the body quickly, and would arguably be a better choice if you were determined to drink your dye. (I want to emphasize that we are NOT advocating this and strongly insist that *you do not try this at home.*) All of these are soluble and lasable in ethanol, so you could, indeed, drink these lasers. (In fact, a droplet of alcohol containing the dye would itself constitute a complete laser, as Hänsch showed. With the light totally internally reflecting from the interior droplet surfaces, the entire laser is edible in principle.) The actual toxicity of even these dyes is not fully established, however. Certainly most laser dyes are toxic, especially the cyanide-based dyes, which have earned them the distinction of an admonitory article in *Laser Focus* in 1975.[4] The toxicity of Coumarin is lower than this, at an LD_{50} of 275 mg/kg.[5]

So laser dyes in gelatine or ethanol are just possibly edible lasers. But we haven't moved very far from the original example. Surely there is some different case?

One that occurs to me is the case of the Color Center Laser. These used point defects in crystal lattices as broad-band, tunable laser sources. Many of these crystals are minimally soluble, indigestible crystals such as Lithium Fluoride, or even diamond. But most of the lasing defects are in Potassium Chloride, or even in Sodium Chloride.[6] Two of the crystals used by the only commercial color center laser were of Potassium Chloride. KCl is certainly edible, when ground up or licked. It's used as a salt substitute. It's true that most of the color centers involve impurities introduced into the crystals, but these are at low concentrations—typically about 1 percent, and most are benign, being other alkali metal ions. But both KCl and NaCl crystals are host to defects that don't involve impurities, such as the F_2^+ laser.

Besides defect lasers, there are also laser materials in which the salt crystal acts as a host medium for lasing molecular ions. My own work on Superoxide ions in salt crystals is just such a case.[7]

Unfortunately, there aren't too many other edible crystals. There are other alkali halide crystals, and other color centers in some of those (such as Rubidium Chloride, or Potassium Bromide), but no one, to my knowledge, has reported laser properties in defects or impurity centers in, say, rock candy, or crystallized proteins. Perhaps this column will spur someone on to take up the challenge.

Having exhausted solids and gases, what about liquids? Here we find our choices limited as well. Water and ethanol are potable, but most other liquids, including organic liquids, are not. You can drink small amounts of things like glycerine, but I haven't been able to lase them, even with inclusions like dyes. Most of what we drink, in fact, is water with other added things. Aside from organic compounds and their perhalogenated analogues, and siloxanes, most substances which are liquid at room temperature are either highly toxic or highly reactive in the presence of organic liquids and water. So what other choices are there?

What about water and alcohol in vapor form? That meets the guidelines—you can't consume them in vapor form, but you can condense and drink them without any other alteration. There are about two dozen lines on which water vapor will lase continuously between 2 and 350 microns.[8] And ethyl alcohol has been found to lase at one wavelength—396 microns. There must be others. Methyl alcohol has over 100 lasing lines in the infrared. One suspects that the researchers weren't looking very hard, so perhaps this is a fruitful research area for future workers. According to D. A. Jennings, K. M. Evenson, and J. J. Jimenez, all working at the NIST laboratory in Boulder, Colorado in 1975,[9] "It is interesting to note that the ethyl alcohol line lased very well on vodka, gin, and rum, but it lased on only one line and rather weakly compared with methyl alcohol. It is quite obvious that there are better uses of ethyl alcohol."

The item that got me interested in the entire topic, however, was the "gin and tonic" laser of which I'd heard rumors ever since grad school. Diligent searching failed to turn up any references in the open literature, so I started asking people about this topic. If they heard of it, it was only as a vague rumor. But finally my queries got back to someone who had been involved in the experiment.

I stress that this was not a serious laboratory experiment, and simply done as a lark. The initial discovery was apparently made by Sam Tuccio at Eastman Kodak Research Laboratories in Rochester, N.Y. in 1968, and Otis Peterson was also involved. He discovered that tonic water could be made to lase when driven by "a very fast flashlamp." The choice of quinine isn't surprising—it's well-known that quinine water and tonic water fluoresce bright blue under ultraviolet excitation, which can make for interesting lecture demonstrations. The peak fluorescence for quinine sulfate is about 450 nm. The fast flash lamp is necessary in order to get a lot of the quinine molecules in their excited state rapidly, since the fluorescence lifetime is about 20 nanoseconds.[10] That the blue fluorescence might also lase was a definite possibility, and the spectral width and short decay time are comparable to those of tunable laser dyes, so that if the quantum efficiency is sufficiently high the quinine solution ought to have sufficient gain to overcome cavity losses. This was evidently the case.

Of course, bubbles in the soda would disturb lasing, and my informant didn't recall how that was dealt with. It would be simple enough to simply let the soda sit out and go "flat." They found that it wasn't a very good laser, however, so it wasn't followed up. Certainly, despite searching the literature, I could find no mentions of a quinine laser. The closest I could find were the laser dyes based on quinolone and azaquinolone, which lase around 400-500 nm, but are structurally not really very similar to quinine.

At the time, Eastman Kodak ran a monthly advertisement in the magazine *Scientific American*. Every month there would be a one-page entry with bits about ongoing research at the Kodak Labs, and in the June 1969 issue on page 51, along with notices about using bacilli to manufacture chemicals and on measuring the reflectance of soils, there was a sort of coy blurb on the tonic laser:

> Omit Gin
>
> A certain well-known brand of quinine water can be made to lase. It lases much better than its equally well-known competitior. We have no intention of disclosing which is which, nor the circumstances that inspired this particular study, nor the precise ultimate objectives of our main study of organic dye lasers. When the work turns up something that ought to be published in a serious way, we present it in a proper forum.
>
> *Meanwhile if some ambitious youngster is asking for your guidance in making a liquid laser, refer him or her to Bruce Burdick. . . .*[11]

Notes

1. T. W. Hänsch, M. Pernier, and A. L. Schawlow, "Laser Action of Dyes in Gelatin," *IEEE Journal of Quantum Elctronics* 7 (1): 45–46, Notes and Lines column (January 1971).
2. Theodor W. Hänsch, "Edible Lasers and Other Delights of the 1970s," *Optics and Photonics News* 16 (2): 14–16 (February 2005).
3. J. G. Kepros, E. M. Eyring, and F. W. Cagle, "Experimental Evidence of an X-Ray Laser," *Proc. Natl. Acad. Sci. USA* 69 (7): 1744–1745 (July 1972).
4. A. Kues and Gerard A. Lutty, "Dyes can be Deadly," *Laser Focus* 11 (5): 59–61 (May 1975).
5. http://en.wikipedia.org/wiki/Coumarin, accessed April 23, 2012.
6. For one example, Joseph F. Pinto, Lawrence W. Stratton, and Clifford R. Pollock, "Stable Color-Center Laser in K-doped NaCl tunable from 1.42 to 1.76 um," *Optics Letters* 10 (8): 384–386 (1985).
7. S. R. Wilk, R. W. Boyd, and K. J. Teegarden, "Laser Characteristics of KNL:O2," *Optics Communications* 47 (6): 404–406 (1983).
8. William S. Benedict, Martin A. Pollack, and W. John Tomlinson, "The Water Vapor Laser," *IEEE J. Quantum Electronics* 5 (2): 108–124 (February 1969).
9. D. A. Jennings, K. M. Evenson, and J. J. Jimenez, "New CO2 Pumped CW Far-Infrared Laser Lines," *IEEE J. Quantum Electronics* 11: 637 (August 1975).
10. David P. Richardson and Raymond Chang, "Lecture Demonstrations of Fluorescence and Phosphorescence," *The Chemical Educator* 12 (4): 279–281 (2007).

11. My original article was cited and expanded on the *Popular Mechanics* website on February 22, 2011, written by Olivia Koski (http://www.popularmechanics.com/science/energy/next-generation/how-to-make-a-laser-from-a-gin-and-tonic, accessed April 23 2012) and in Olivia Koski's *Physical Science Blog* on *scienceline* on April 4, 2010. http://scienceline.org/2010/04/drinking-and-lasing/, accessed April 23, 2012. Ms. Koski contacted Mr. Burdick, who didn't remember much about the process, but said that the mini-article was "meant to generate buzz about lasers," which still weren't very familiar to the general public.

{ 28 }

Pyrotechnic Lasers

Lasers have been pumped using flash lamps, arc lamps, sunlight, and other lasers—but what about pyrotechnics? This possibility has intrigued me ever since I started studying lasers. Burning metal, such as magnesium, is extremely bright and has a broad spectrum with a lot of hard ultraviolet light, which is excellent for pumping many visible and near-infrared lasers. Moreover, a pyrotechnic source requires no warm-up time, no cooling systems, and no supporting electrical systems. Transfer is very efficient, and the light pump is ready to go at a moment's notice.

Of course, the considerable downsides are that the source is short-lived; that pyrotechnic decay products—smoke and ash—interfere with pumping; and that the resulting high temperatures and detritus of combustion mean that the source will either destroy the laser or must be contained in a cartridge that could be removed and disposed of. Pyrotechnic pumping, one would expect, would serve only niche applications.

Over the years, I'd heard hints of pyrotechnic lasers, but they were mentioned only rarely, and they were virtually never discussed in public forums. Nevertheless, it seems that, at least to some researchers, the topic was of interest right from the start. W. E. Bushor,[1] in a 1962 issue of *Electronics*, reported on the use of light from exploding wires to pump lasers. And in the 1963 report,[2] called *An Investigation into the Feasibility of a Pyrotechnic Laser Pump*, Chester L. Smith and Paul Kisatsky of the Picatinny Arsenal identified several promising formulations. They mentioned zirconium metal with potassium chlorate as an oxidizer as the brightest source. One of the authors went on to build such a pyrotechnically pumped laser using a cyanogen-oxygen mixture that was ignited to produce a 1- to 3-ms light pulse to drive a neodymium-glass laser in one set of experiments and a ruby laser in another.

On December 29, 1964, William Buchmann was granted patent No. 3163799 for "Exploding Squirted Wire Pulsed Light Source," explicitly for pumping a laser or optical maser. It was assigned to Hughes.

In an article in *Applied Optics* in 1967, researchers reported a 140-ms pulse for the ruby laser at 693 nm. One individual, who seems not to be associated with any large organization, obtained three patents for pyrotechnically pumped lasers between 1966 and 1972 (Nos. 3271696, 3309620, and 3646471). He says that the devices "provide a laser light generator which is portable, storable, reloadable, and adapted to uses for and within pre-selected portions of the ultraviolet, visible, and infrared regions of the spectrum." He goes on to state that "an object of this invention is to provide a laser light beam generator for research, educational, therapeutic, and like associations." This is laudable and benign, and one can't help observing that he doesn't include military or defense applications. Many of the pyrotechnic pump laser patents are, in fact, assigned to defense contractors, and are evidently intended for use as weapons themselves, as target designators, or (in some cases) initiators for simultaneous detonation of explosives.

A. A. Kaminskii and co-workers at the Moscow Institute of Crystallography took Smith and Kisatsky's Zr:KClO4 mixture and used it as the pump for 29 rare earth lasers in the range of 1.037 to 2.36 μm.[3-5] Kaminskii continued to pursue pyrotechnically pumped lasers years later, with publications in the *Soviet Journal of Quantum Electronics* in 1983 and *Laser Physics Letters* in 2006.

Dozens of US patents have been granted for pyrotechnic pumping since the early 1960s, although no one knows how many of these sources have been built. In one of the ones that was constructed, Claude R. Jones et al. used exploding metal films to drive molecular iodine at 342 nm and atomic iodine at 1.315 μm.[6]

In 1981, W. F. Wing et al. of Sanders Associates reported lasing of Nd:YAG rods with a pump of zirconium in oxygen.[7] They achieved continuous-wave output for 2 seconds and average power of 5 Watts. Although they predicted that even longer output times could be realized by scaling up the system, I have found no reports verifying this. About the same time, GTE Labs in Massachusetts was pursuing percussively ignited flash lamps for, among other things, laser pumping.[8-10]

One use for such pyrotechnic lasers is as a source for illumination and laser radar, especially for systems used in satellites. Northrop Grumman tested such a device, although the type of laser and pump were not specified in their presentation about it at the 1986 IEEE Aerospace Applications Conference. Schwartz Electro-Optics tested one as well, as they report in *Proceedings of SPIE* in 1992—a Nd:Cr:GSGG composite rod lasing at 1 and 3 μm, using zirconium and oxygen as the pump.

A patent from 1983, titled "Low Cost Laser,"[11] points out another potential purpose for pyrotechnic lasers: Pyrotechnics could provide a lightweight, compact mechanism for powering a hand-held laser gun. Two more patents for such a hand-held laser gun were issued in 1997 and 1998.[12]

More recently, in *Chinese Optics Letters* in 2008,[13] Nan Xiao and co-workers at the National University of Defense Technology in China have used pyrotechnic mixtures of their own devising to pump neodymium glass rods, producing an output of 5.5 W at 1.053 μm. They don't state the intended purpose of this work.

Clearly, many relevant papers and patents emanate from defense-oriented companies and institutions—and more still may have been built of their own devising to be proprietary.[14-16] Besides the resources that I've already cited, news reports have noted a military interest in the pyrotechnic pumping of lasers, probably due to the lightness and portability of the pumping scheme.

I'm surprised that I've found no reports among amateur experimenters, for whom this would represent an easily accessible high-intensity pump. But I suspect that the possibility of destroying the rod and mirrors—all of which are not cheap to replace—must outweigh any such advantage. After all, no one wants to see their scientific investment go up in smoke.

Notes

1. W. E. Bushor, "Sun and Exploding Wires Pump Lasers," *Electronics* 35 (13): 24–25 (1962).

2. The Smith and Kisatsky report is DTIC document 420238, Technical Report 3102 from Pickatinny Arsenal, Dover, NJ, August 1963, available online at: http://www.dtic.mil/cgi-bin/GetTRDoc?AD=AD0420238, accessed April 29 2013; C. L. Smith, E. Homentowski, and C. Stokes, "Direct Nondestructive Pumping of Ruby Lasers by Chemical Means," *Appl. Opt.* 6(6): 1130–1131 (1967).

3. A. A. Kaminskii et al., "Stimulated Radiation from $Y_3Al_5O_{12}$ – Nd^{3+} Crystals," *Soviet Physics JETP* 24 (1): 33–39 (1967).

4. A. A. Kaminskii et al., "Pyrotechnically Excited Quasi-CW Laser," *Soviet Physics—Technical Physics* 14 (3): 396–402 (1969).

5. A. A. Kaminskii et al., "New Quasi-CW Pyrotechnically Pumped Crystal Lasers," *Sov. J. Quantum Electron.* 13 (7): 975–976 (1983).

6. C. R. Jones and K. D. Ware, "Optically Pumped Ultraviolet and Infrared Lasers Driven by Exploding Metal Films and Wires," Los Alamos Conference on Optics 1983 *Proc. SPIE* 380: 60 (1983) and U.S. Patent No. 4,599,731 (1986).

7. W. Wing, R. Grasso, J. Mosto, S. Schecht, J. Baer, and E. Chicklis "Pyrotechnically Pumped Laser," *IEEE Journal of Quantum Electronics* 17 (12) 2360 (December 1981); P. Pencikowski and P. Csik, "A Long-Range Synthetic Vision System Combining a Pyrotechnic-Pumped Laser and a Range-Gated Camera," *IEEE Aero-space Applications Conference*, Snowmass, Colo., USA, February 1996, 97–102 (1996).

8. John F. Waymouth and John W. Shaffer, "Investigation of the Feasibility of Pumping High Energy Neodymium Glass Lasers by Chemical Flashlamps." DE85-012962 (1984), http://www.ntis.gov/search/product.aspx?ABBR=DE85012962, accessed September 22, 2012.

9. J. F. Waymouth, A. C. Bouchard, R. A. Fowler... *Method of making a percussively-ignited photoflash lamp*—US Patent No. 4,278,310 (July 14, 1981).

10. J. F. Waymouth, A. C. Bouchard, R. A. Fowler... *Percussively ignited photoflash lamp and method of making same*—US Patent No. 4,201,540 (May 6, 1980).

11. Evan P. Chiklis and James R. Mosto, *Low Cost Laser*—US Patent No. 4,371,969 (February 1, 1983).

12. William R. Houde-Walter, *Laser Gun and Cartridge*—US Patent No. 5,617,444 (April 1, 1997) and *Laser Gun and Cartridge* US Patent No. 5,761,235 (June 2, 1998).

13. N. Xiao et al., " ~mJ Long Pulse Pyrotechnically Pumped Laser," *Chinese Optics Letters* 6 (8): 578–579 (2008).

14. M. Acharekar and R. LeBeau, "Miniature Laser Direct-Detection Radar," Laser Radar VII: Advanced Technology for Applications *Proc. SPIE* 1633: 94–111 (1992).

15. A. A. Kaminskii et al., “5.5 J Pyrotechnically Pumped $Nd^{3+}:Y_3Al_5O_{12}$ Ceramic Laser,” *Laser Phys. Lett* 3 (3): 124–128 (2006).

16. A recent Chinese paper appears to be a history of pyrotechnic lasers. Xiao Nan, Jiang Zongfu, Hua Weihong, and Yuan Shengfu, “Investigation Development of Pyrotechnically Pumped Laser,” *Laser & Optoelectronics Progress* 46 (3): 32–43 (2009).

{ 29 }

Defunct Lasers

I don't think that anyone has ever done an accurate count of the number of lasers that have been invented. Wikipedia[1] says merely "Many Thousands." Marvin J. Weber's *Handbook of Laser Wavelengths* lists over 15,000 different lines.[2] While many lasers have more than one line, and some lasers are not represented even in this comprehensive volume, that's probably a good estimate. Some 80 percent of these are gas lasers.

Contrast this with the very small number of lasers in common use. Wikipedia lists about 50 lasers on its "Types of Lasers" page. Weber lists about 150 in his "commercial laser" section. The numbers are a bit low, however, since they group all Dye Lasers together as one. And Exciton, to give just one manufacturer, lists some 300 commercial dyes on its webpage.[3] There are similarly many types of semiconductor diode lasers, and if one allows for custom tailoring to meet specific wavelengths and other properties, there are a great many variations of these. There are probably "off-the-shelf," something like 1,000 lasers available in large quantity, with probably only a couple of hundred at most dominating the field.

Still, most of the lasers in use, even in laboratories, are probably the same few—HeNe Nd:YAG, a few diode lasers for direct use or for pumping, CO_2 lasers, Ti: Sapphire, Krypton, and Argon lasers, excimer lasers, a few dye lasers, and so forth. In a sense, therefore, *most* lasers are "defunct" lasers, not commercially available.

Inevitably, most of those 15,000 lasers are not going to be used. There simply isn't a market for every conceivable laser. The history of the development of a particular laser, the economics of its production, its ability to fill a much-needed niche market (or its operation in a regime already well-covered by other lasers) all factor into whether a laser will be manufactured by one or more suppliers. Other lasers do continue to be made in the laboratory to fulfill particular needs, even if it does not make commercial sense to market such a laser. But, with so many potential sources, most of them will be unknown to researchers unless they make an effort to seek them out. And, in fact, many of those 15,000 lasers received no more than a single mention in research papers exploring the types and properties of lasing materials. It's likely that not enough is known about them to tell whether they'd be a good choice as a laser for any given application.

Sometimes it's pretty clear why a laser isn't being used—factors such as cost, convenience, special handling requirements, and the like can determine whether a laser is favored or not. It's surely easier to use a rugged solid-state laser that merely has to have a cooling system switched on, for instance, than to use a laser that requires an oven to heat the laser material to high temperatures, or a cryogenic system to maintain it at low temperatures, or one that uses fragile glass tubes to hold a lasing gas. I suspect the last of these (coupled with the extension of the lasing wavelength range in recent years) is the reason for the preference for solid state over gas lasers in both civilian and military applications. That doesn't bode well for that 80 percent of the possible lasers that use the gas phase.

To take an example from my own experience, Weber lists about 60 Color Center Lasers, which use as their gain medium point defects in crystals. The advantage of such lasers is that they lase over a broad range of wavelengths in the near infrared, where dye lasers tend to be too ephemeral. They can be tuned for spectroscopy or used for short pulses. There has only been one commercial manufacturer of these lasers, however, and they only produced four of the possible sixty. Today no one produces them, and any such lasers in use are built in the laboratory. Why is this?

Most color centers can have their electrons easily excited into the conduction band of the host crystal by a combination of optical and thermal excitation, and the liberated electrons eventually bind to a non-color-center site, so eventually enough of the centers are lost to lasing unless the crystal is stored at low temperatures (usually liquid nitrogen temperature, although sometimes lower) or in the dark, or both. Operation usually requires cooling to the temperature of liquid nitrogen or lower, and thus requires a reliable vacuum system with cold traps. If the crystal warms up, it can sometimes be rehabilitated, but often the lasing crystal must be replaced. Although the first operation of a color center laser was done with flash lamp excitation, and some have been diode pumped, most such lasers have required excitation with another laser. In many cases you can generate a beam of the required wavelength with fewer restrictions using another method, for instance, using an Optical Parametric Oscillator.

Similarly, some of the first lasers built, trivalent Uranium or Samarium in a CaF_2 host, require liquid helium temperatures for operation.

On the opposite end of the temperature scale, metal vapor lasers require high temperatures to maintain the required vapor for metals such as copper (typically 1500°C), gold (1650°C), or lead (1100°C).[4] Many of these don't seem to be manufactured commercially in the United States anymore, although they are manufactured outside the United States for dermatological applications.[5]

Other lasers use expensive or hard-to-work materials. The H_3 center in diamond, lasing at 530 nm, requires an appropriately treated diamond host. Most of us have little use for nuclear-pumped lasers.[6] The problem of thermal effects in neodymium lasers was minimized by going from a solid host to liquid hosts. But most of these host liquids were aprotic acids that were corrosive to the container and potentially dangerous to the human operators. Interest in this type has revived in laboratories now that there are efficient diode pumps, but I know of no commercial systems using this medium, although there are a great many solid-state neodymium lasers.[7] Then there are the lasers which have been discovered but not even officially published. As we point out elsewhere in this book, quinine dissolved in soda has been shown to lase, but no one has pursued it to the point of publishing a paper on the topic. An examination of the literature shows no papers by anyone on the subject. Similarly, neodymium ions may be dissolved in water, and if one substitutes heavy water one may minimize the absorption by the OH^- tail that reduces the gain, resulting in a liquid neodymium laser using a much less toxic host. I have been told of such experiments, but the results have gone unpublished, since the gain cross-section was too low.

Some lasers have gone unused for reasons that are not altogether clear. For instance, there are a great many possible metal vapor lasers. In October 1980 *The Amateur Scientist* column in *Scientific American* even published a detailed article on how to build your own. It was later reprinted in one of their books, and eventually found its way on-line. The HeHg laser can produce lines at 567 and 615 nm. It's easy enough to build that a home hobbyist can do it. The mercury is no more toxic than that in a fluorescent tube and the high voltages are comparable to other gas lasers, like the ubiquitous HeNe. So why has this laser never been developed commercially? Certainly the hobbyists want to know.[8] The relatively long yet small diameter laser tubes probably explain the lack of commercial interest in this laser. It's also possible that at first it was thought to be too close to existing lasers, or to have no purpose worth developing, and that now the preference against gas lasers would prevent any adoption.

Since I originally published this paper, I received a letter from a respected laser scientist commenting on it. He protested the picture with the gravestone saying "Metal Vapor Lasers" and said that they definitely were not dead yet. In my defense on that, I do not do the illustrations, although I did approve its use. I didn't think it would be taken so literally. The writer further observed that there were several on the market, and that high temperatures were not at all necessary for all of them.

Notes

1. http://en.wikipedia.org/wiki/List_of_laser_types, accessed September 2012.
2. Marvin J. Weber (ed.), *Handbook of Laser Wavelengths* (1999).
3. http://www.exciton.com/, accessed September 30, 2012.
4. William T. Silfvast, *Laser Fundamentals*, 2d ed. (2004), 505.
5. http://www.yachroma.com/auran/, accessed September 30, 2012.
6. Weber, *Handbook of Laser Wavelengths*, 719–723.
7. http://www.laserfocusworld.com/display_article/201561/12/none/none/News/Flowing-neodymium-offers-improved-heat-management, accessed September 30, 2012.
8. http://www.repairfaq.org/sam/laserchg.htm, accessed September 30, 2012.

{30}

The Phantom Laser

If you've built or aligned a laser, you know how difficult the task of alignment can be. It's a standard procedure for many multi-line systems, such as Argon Ion or Krypton Ion lasers to perform alignment to maximize a particular line. In many cases setting the proper alignment is a simple hunt through the two-space Sea of Tip and Tilt for the Island of Maximum Output. But in many cases optimum alignment requires more than merely the adjustment of one mirror. Mirror spacing must be properly set, the cavity axis must coincide with the axis of the gain medium, and the beam waist must be properly located relative to the gain medium. In many cases the laser output won't be sufficient if these conditions aren't fulfilled.

I've aligned a lot of lasers in my career. For my doctoral work I had to construct a laser cavity around the sample under study for every experiment. The procedure became routine after much practice, but it was still long and involved, and it was never trivial. Lasers don't simply fall into alignment of their own accord. Usually.

There *are* a few cases of self-organizing laser resonators. The "whispering gallery" modes of microspheres (droplets of solvent containing rhodamine 6-G, for instance, or solid glass microspheres containing some gain medium).[1] Semiconductor lasers whose cleaved ends are dictated by crystal structure require no further adjustment. Amplified Spontaneous Emission (ASE) in long samples of gain medium (such as in many nitrogen lasers) doesn't really constitute a resonator, but people call them "lasers" by courtesy. Natural "lasers" of this sort can occur on astronomical scales.[2]

But in most cases laser alignment is a careful and time-consuming process. That's why I was really annoyed when I found that an Unstable Resonator system I was working on spontaneously decided to become a perfectly-aligned Stable Resonator. The power concentration at the end mirror became so great that it blew the coating off the multi-layer stack. After all my years of struggling with laser alignment, how is it that the one time I really *don't* want it, I get a perfect alignment that destroys my mirrors? And how did it happen? The parts I was using *shouldn't* have been capable of forming a stable resonating cavity.

It was an unstable resonator, essentially a 3:1 telescope with one convex end mirror and one concave end mirror having three times the focal length, placed so that their focal points coincided. If you follow a beam in one direction and it gets smaller by a factor of three with each round trip, until diffraction effects begin to dominate and you get an unchanging "diffractive core." Trace the beam in the other direction and it gets larger with each pass by a factor of three. It would simply expand until it was larger than the end mirrors, were not the usual procedure to remove the outside of the beam during its last pass using a "beam scraper." This is typically a mirror mounted at 45° to the beam axis. It has a square or circular hole cut in the center to allow most of the beam to pass through, but the outer portion is redirected out of the resonator path to serve as useful output. It's useful output with a hole in the center, of course, but over a sufficient distance from the laser diffraction fills in the hole, albeit irregularly.

In performing my "post mortem" on the mirrors, I was struck by the fact that the line that had been burned into the coating was cleanly horizontal. Under circumstances most of us are familiar with, photons don't care about gravity, so the fact that the resonator produced a high intensity horizontal line focus on one mirror was a valuable clue. Clearly I should look for structures with that symmetry. The horizontal and vertical walls *were* different—the electrodes used to excite the discharge were on one set—but not different enough to cause this effect. Furthermore, reflections from the electrodes or the walls weren't supposed to happen. There were spoilers in place. Lastly, even if reflections from the wall were contributing, wouldn't the resulting mode be wider than a thin horizontal line?

Aside from the walls, the only other item that had that symmetry was the Beam Scraper. It was square, with a square hole cut into it to allow the beam to pass, and one of the sides was parallel to the horizontal damage line burned into the end mirror. But how could the scraper mirror cause that effect? It's on the outside of the diffraction core.

Of course, it was something simple, and easily overlooked. The Scraper Mirror was usually drawn in schematics as a thin mirror at a 45° angle, floating miraculously in mid-beam. In the real world, the mirror has to be supported. It was, in fact, bolted to the same platform as the End Mirror, and it slipped over the End Mirror Mount. The square hole bored through the center had perfectly straight sides parallel to the Optical Axis. They weren't polished, but at glancing incidence, they made perfect mirrors.

The combination of the long-radius end mirror with the walls of the bore through the scraper mirror very nearly made up a corner cube, retroreflecting the beam back nearly anti-parallel to its original course.

But it wasn't a perfect corner cube, for two reasons. One is that the end mirror isn't flat, but is curved. That means that beams that lie in planes not running through the center of the mirror *and* not perpendicular to the sides of the bore will eventually walk off the axis. In other words, they're not stable, and don't eventually retrace their paths. We should expect stable resonating only along a thin horizontal line and along a thin vertical line, both passing through the optical center.

The other reason the combination isn't a perfect corner cube is that the bore through the scraper mirror isn't the same length everywhere. The scraper makes a 45° angle with the optical axis, facing "downwards." So the bore is longest at the top and shortest at the bottom. This means that you get a lot of returns from the "corner cube" formed by the top surface of the bore and the end mirror, but far fewer from the "corner cube" formed by the lower wall of the bore and the end mirror. Too many ray paths that start out looking stable at the top end up leaking out through the bottom. The horizontal "corner cubes," however, formed by the sides of the square borehole through the scraper mirror and the end mirror, were both the same length. Apparently these quasi-corner cubes returned enough of the rays to allow a stable resonator to build up along the horizontal direction, but the short length of the borehole along the bottom prevented a stable resonator from existing along the vertical axis. So only one stable resonator existed, formed by the combination of the long radius convex mirror and the vertical walls of the borehole though the scraper mirror on one end, and by the concave mirror on the other end. But only along the plane that ran through the optical axis.

Once the source of the problem had been surmised, a solution could be suggested. In this case, the solution was obvious—cut back the walls of the square borehole so that they were no longer parallel to the axis. If they flared outwards from the scraper mirror end they wouldn't form quasi-corner cubes, and the reflected rays would not be reflected back onto stable paths. So that is what we did. And after that, there were no more horizontal burns to the coated end mirror.

Notes

1. There is a vast literature associated with these. For recent papers, see http://www.opticsexpress.org/abstract.cfm?URI=OPEX-8-11-605, accessed September 22, 2012, or V. Sandoghdar, et al., "Very Low Threshold Whispering-Gallery-Mode Microsphere Laser," *Phys. Rev. A* 54: R1777–R1780 (1996). A brief introduction is at http://metrology.hut.fi/courses/s108-j/Nano2.pdf, accessed September 22, 2012.

2. See, for instance, http://www.ast.cam.ac.uk/HST/press/laserils.html, http://home.achilles.net/~jtalbot/news/MWC349.html, http://adsabs.harvard.edu/cgi-bin/bib_query?1995A%26A...300..843T, http://home.achilles.net/~jtalbot/history/stellar.html, and http://home.achilles.net/~jtalbot/news/EtaCarinae.html, accessed September 22, 2012.

In print, C. Thum, V. S. Strelnitski, and J. Martin-Pintado, "Hydrogen Recombination {beta}-lines in MWC 349," *Astronomy and Astrophysics* 300: 843 (1995); and R. C. Doel, M. D. Gray, D. Field, and K. N. Jones, "FIR Lasers from Dense OH Maser Regions," *Astronomy and Astrophysics* 280: 592 (1993).

{ 31 }

The Case of the Oily Mirrors

A LOCKED ROOM MYSTERY

Every physicist learns that there are a few Great Solved Problems in physics—the Harmonic Oscillator, the Particle in an Inverse Square Force, The Symmetric Top—so, too, there are the Great Mystery situations. One of the classics is The Locked Room, in which a crime takes place within a sealed room, with no way for the criminal to get in or out, apparently. Arthur Conan Doyle arguably invented the scenario in his second Sherlock Holmes novel, *The Sign of Four*, in which Bartholomew Sholto is found murdered by a poisoned dart in his sealed study at Pondicherry Lodge. The gimmick has been used countless times since, and the trick is rarely anything so obvious as "here's a trap door you didn't know about"—it's usually "here's something you overlooked." A good mystery, like a good science question, turns upon new ways of interpreting data, leading to greater understanding.

I had my own Locked Room Mystery once. I was working with a rather tight arrangement of optical components. They were mounted on a vertical bench that was intended to be packed inside a closed space. We had a very small but elegant CO_2 laser, several copper mirrors for directing and focusing the beam, beamsplitters, and so forth, all tightly tied down in a sturdy configuration.

After the laser had been operating for some time, however, I noticed that the beam, when visualized with heat-sensitive paper or liquid crystal sheets, was starting to look diffused, instead of the nice, tight beam we'd started out with. Nothing had changed at the laser itself. But the beam was certainly degrading.

Finally, I noticed that the copper mirrors being used to steer the beam were looking sort of fuzzy, as if the specular surface was turning to a matte finish. I dismounted a couple of them and took them off to the microscope. Lo and Behold! The surfaces of the first mirror after the laser, and the one after that, were covered with very fine drops of some liquid.

Since the temperatures of these mirrors had been kept pretty constant, it didn't seem that the cause could be condensation. What's more, the liquid didn't evaporate with time. I could only imagine that this non-volatile liquid was some sort of oil.

But where was it coming from? That was the real mystery. This optical setup was not merely in a laboratory, but in a Clean Room, from which such notorious oil sources as vacuum pumps had been banned. The walls of the room had been painted with a special compound to trap dust, and we had to walk over renewable sheets of sticky material that lifted dust from our shoes. Actually, it wasn't a full Clean Room that demanded suiting-up and hairnets, so perhaps I should say it was a Not-So-Dirty Room. Nevertheless, the precautions were sufficient to prevent contamination by liquids unknown. Nobody was opening cans of Coke where they could spray on the lab mirrors.

Even more confusing, the droplets weren't uniformly distributed over the surfaces of the mirrors. They were larger and more numerous where the beam was more intense. As the laser was running in TEM_{00} mode, the droplets had a Gaussian distribution in size and number. So there were *two* mysteries—where was the oil coming from, and why was it distributed this way? Nobody knew.

These two phenomena *had* to be connected, so it seemed to me that maybe the laser was heating up the mirror and causing the mirror to "sweat" oil from microscopic pits, maybe at the grain boundaries. I mentioned this possibility to a colleague, who said he'd heard that some copper mirrors were polished under olive oil (of all things). It sounded weird but promising, so I looked into it. And I found that there weren't any of the hypothetical microscopic pits, and that these mirrors certainly weren't polished under *any* oil. I confirmed this by talking to the technician who had polished our mirrors.

So, back to Square One. If it wasn't the heat from the laser, then what *could* have produced the drops? What other factor correlated with the beam profile? Then it struck me. Optical Levitation.

I'd read an article about Optical Levitation in *Scientific American*[1] many years before and had looked up the original papers.[2, 3] You can suspend a glass microsphere against the force of gravity with an upward-directed laser beam, in just the way that a jet of water can push a ball upwards. The laser beam differs from the jet of water in that it actually provides a *stable* support. If the microsphere wanders off the exact center of the beam, it is pushed back toward the axis, while a jet of water pushes the ball further from the axis. (The force is also pretty small, so the glass microsphere is only about 20 microns across.)

Optical Levitation can be viewed in the following classical way: The incident ElectroMagnetic wave stirs up currents in the object they interact with. These currents in turn, interact with the original EM wave, resulting in a force on the object. The effect was first predicted using thermodynamics. But it is perhaps more intuitive to view it quantum mechanically as due to the transfer of momentum from photons (each with momentum $\mathbf{h\upsilon}$) to the object. In many cases of physical interest the material is a dielectric, with only part of the beam reflected, and with part refracted. In fact, that's what keeps the microsphere stably supported in the center of the laser beam—the beam is refracted upon entering and leaving the sphere, and some of the momentum transferred to the microsphere is sideways as well as upward. The imbalance between the sideways momentum at the side closer to the beam axis and the side away from the beam axis tends to push the sphere back to the center. (You can perform this trick with perfectly reflecting metal microspheres, too, but you have to suspend them in a beam having a TEM_{01^*} "donut" mode, for obvious reasons.)

The same forces that keep a carefully balanced microsphere stably on the beam axis will also tend to push randomly positioned floating dielectric spheres *toward* the beam axis. Aim a powerful laser beam through a fine cloud of particles and you have a "broom" that will tend to sweep those particles forward and toward the beam axis. Fine droplets of oil form spheres as surface tension acts to minimize surface area, and they can be suspended in air for quite a long time.

That seemed like a plausible explanation for our oily mirrors—the CO_2 beam was passing through a cloud of minute oil droplets, grabbing them, and pushing them toward the beam axis. Eventually the droplets landed on the copper mirrors, where they stuck. And they'd tend to cluster near the axis of the beam. You'd even get buildup on the second mirror, until the buildup on the first mirror distorted the beam so much that it was too diffuse to push the droplets toward the second mirror. You'd expect the first mirror to have more buildup on it, and that's just what we saw. Bingo!

But where was the cloud of oil droplets coming from? This was supposed to be a Clean Room. There was nothing to throw off the offending cloud of droplets. No pumps in the room. Nothing lubricated (not even the tip/tilt mounts). No aerosol generators. All we had was the optical bench, the optical components, and their mounts. There was a beam block that we used to put in front of the laser when we worked on the optics, but it was made of inert material.

The beam block. It may have been inert, but I learned that, to minimize particulates, it was cut *under oil*. Some may have diffused into it or gotten trapped on the surface, even though it was apparently clean and dry to the touch. Then, when the beam block was hit by the laser beam it would heat up, and some of the oil would vaporize, recondensing in the cooler air as droplets.

The solution: replace the beam block with one that has never seen oil. The result: The oil droplets went away. Case Closed. And Locked.

Notes

1. A. Ashkin, "The Pressure of Laser Light," *Scientific American* (February 1972).

2. A. Ashkin, "Acceleration and Trapping of Particles by Radiation Pressure," *Phys. Rev. Lett.* 24: 156ff. (1970).

3. A. Ashkin and J. M. Dziedzic, "Optical Levitation by Radiation Pressure," *Appl. Phys. Lett.* 19: 283ff. (1971).

{ 32 }

Pinhole Glasses

If you type the words *Pinhole Glasses* into an Internet search engine, you'll be introduced to hundreds of sites offering to sell you "wonderful" devices for helping you see that will work regardless of your prescription, yet which cost less than $20. You can also find them listed as *stenopeic glasses* or *dioptric pinhole grid glasses*. Some are sold through well-known commercial sites, such as Amazon, while others are on pages associated with Alternative Medicine, such as the Bates Eye Method, but most are from sites that are simply selling the glasses for profit.

These sites generally claim that these low-tech glasses provide an inexpensive alternative to conventional glasses for any prescription, although they acknowledge that the glasses take some getting used to. They observe that these glasses have been used outside the United States for many years, and some hint darkly that the reason these are not more widely used is because eyeglass manufacturers and ophthalmologists have conspired to keep them off the market.

There is an air of something clandestine and suspicious about these glasses. Surely there is no Grand Conspiracy to keep affordable technology out of the hands of people, and the glasses look so cheaply made. So what is the real story?

In the first place, one should note that the glasses actually do work. As anyone who has played with pinholes and apertures has found out, you really can see better detail while looking through a single pinhole. Those of us with myopia have found that you can see detail and eliminate the blur of distant scenes by viewing through a pinhole. This comes at the cost of throughput—pinholes severely restrict the amount of incoming light—and of the range you can see without turning your head.

The pinhole here is doing the same thing that one naturally does when trying to make things clearer—it provides the same aperture limitation as a squint. But the pinhole is an effortless squint, and it's of far better quality. I suggest elsewhere in this book (see chapter 1, "Ancient Optics") that pinholes might have been used in the ancient world as aids for magnification. But they might also have been used as a substitute for eyeglasses. The technology to produce a corrective lens is complex and would be expensive—getting clear rock crystal or bubble-free and inclusion-free uncolored glass, then the materials and skill to grind and polish a lens would be a complex undertaking in Classical Greece or Rome. It's far easier to punch or drill a hole in a piece of leather, wood, stone, pottery, shell, bone, parchment, or other material. It could be worn monocle-style between brow and cheekbone, or held before the eye lorgnette-style. It might even be worn as glasses, although this seems less likely. And it's entirely possible that some of these devices may have survived into the present day, and simply aren't recognized for what they are. William Hanks Levy,[1] in *Blindness and the Blind: A Treatise on Typography* (1872) observes that "the ancients had no knowledge of glasses for assisting impaired sight, we may conclude from their universal silence on this matter." But he goes on to note that "We are told that old men among the classical ancients read through a simple tube, which, by isolating objects, made vision more distinct." Levy is here citing *The Generation of Animals*, attributed to Aristotle,[2] which states, "The man who shades his eye with his hand or looks through a tube will not distinguish any more or any less the differences in colours, but he will see farther" and "distant objects would be seen best of all if there were a sort of continuous tube extending straight from the sight to that which is seen." I have experimented with soda straws and find that they do cut down on many of the rays that cause blurring, although they do not function as efficiently as a pinhole.

If you punch a number of holes in your low-tech glasses, as the modern pinhole glasses do, then you will not only increase the amount of light you let in, but will also cover the range of vision that you can see without moving your head. The cost in this case is a multiplicity of images, which takes some getting used to, but can be done. Were I a stranded time traveler in Pliny's Rome, I'd make a pair of such multi-pinhole glasses as spares. In the classic *Twilight Zone* episode "All the Time in the World," nuclear holocaust survivor Burgess Meredith takes solace in the thought that he finally has the time and leisure to read all those books he wanted—until he breaks his idiosyncratic corrective glasses, which no one is left to repair. But he could have remedied his own situation by making his own pinhole glasses as a low-tech solution. Desert Island survivors could use these, as well. The same suggestion was made by "T.W."[3] of St. Cloud, Minnesota, in the June 2007 issue of *Backpacker* Magazine. The nearsighted hiker, worried about losing or breaking glasses, suggests using one's knife or needle to punch multiple holes in a piece of cloth or duct tape to act as temporary emergency glasses "It's not perfect, but it's better than walking off a cliff."

There is some evidence that such pinhole glasses were used in the past, although the question is somewhat tangled. The earliest unequivocal reference I've found to the use of pinhole viewers in the West is a reference to masks with holes in them, made by Paulus Aeginēta, a seventh-century physician and medical writer in his *De Re Medica Libri Septum* (see note 8 in chapter 1, "Ancient Optics")

Pinhole glasses made of metal sheeting with multiple pinholes that could ride atop the bridge of the nose were used in India, according to pediatric ophthalmologist Lea Hyvärinen.[4] Dr. Hyvärinen reports that a pair of these reside in the Museum of Medical History in Helsinki and are believed to be a millennium old.

Her works give no provenance for these items or reason for dating them as being this old.

Moreover, the metal spectacles so closely resemble the items found in Chinese T'ang dynasty tombs (eighth century C.E.) from the region of Astana in Turfan, the Xinjiang Uygur Autonomous region, that I suspect her source has either misidentified these items, or she is mistaken herself.

In 1967 a tomb was opened there, and within the tomb was found an elliptical "eye shield" measuring 17 cm horizontally, 4 cm vertically, and 3 mm thick. It is rimmed with silk, and has an indentation for the nose, so that it may be worn like a pair of spectacles. There are many tiny perforations in the vicinity of the eyes.[5] The general consensus seems to be that these are either burial clothing, not intended for daily use, or that these are "snow goggles" for restricting the overall amount of light entering the eye, similar to Inuit and Nenets (Russian Samoyed people) "snow goggles" make of ivory, bone, or wood or (in the Nenets case) of metal. Snow goggles are a special case. They consist of an eye covering with a horizontal slit, one for each eye. Except in rare cases, the opening is too large even in a vertical direction to serve as a "pinhole" and don't appear to be useful for aiding vision in that fashion. In any event, the slits are very wide from side to side, allowing vision over a wide region, but not usefully restricting rays that might cause blurring.

The Astana metal "spectacles," on the other hand, are very different. They are made of metal (bronze and lead "glasses" have been found) and consist of two "lobes," one covering each eye, which are pierced not by horizontal slits, but by a collection of holes in the vicinity of the eyes. This allows vision over a range, but the apertures are all relatively small in size, no more than a couple of millimeters at most. This is still a useful size for myopia correction, although smaller pinholes produce a better effect. On the other hand, larger holes like this let in more light. It is true that some such metal "glasses" have their holes arranged to form the outline of an eye and a horizontal "lid," which might suggest that they were ceremonial, but these would function perfectly, as well. Other cases seem to have the holes in a random fashion, which is not aesthetically pleasing for a burial costume, but functions very well indeed. On balance, it seems to me very likely that these were functional eyewear, providing relief from myopia while letting in more light than tiny pinholes (good if you're not trying to read distant letters, but just want to see), while also providing some relief from wind, dust, and snow blindness. That they were found in tombs might mean no more than that the deceased were buried with the accoutrements necessary for everyday life.

One of the first appearances in print of pinhole array glasses seems to be US Patent No. 564,518, granted on July 21, 1896, to Franz Heilborn "a subject of the King of Prussia, Emperor of Germany, and a resident of Breslau, in the province of Silesia, Kingdom of Prussia, Empire of Germany." The patent title is simply "Eyeglasses," and Heilborn clearly states that these glasses would be an aid to vision for people afflicted with extreme cases of myopia or astigmatism, for whom standard glasses must be extremely thick, and yet would still create distorted images. Heilborn's pinhole glasses would not. They consisted of sheets of brass (just as in Xinjiang!) or "German Silver" (an alloy of copper, nickel, and zinc, often used for tableware) with a radial or concentric array of "perforations" attached to a standard set of glasses with a hinge. This allowed the perforated plate to be swung in front of the glass, or away from it, giving both standard correction and pinhole correction, as desired.

Many years later Charles Claude Guthrie,[6] M.D., Ph.D., Professor of Physiology and Pharmacology at Washington University Medical School in St. Louis, obtained Patent No. 1,959,915, granted May 22, 1834, on "Lensless Spectacles." He also published an article on "Physiologic Lensless Spectacles" in the *Archives of Ophthalmology* at the same time.

Guthrie's spectacles, as the title implies, used no refractive elements at all, relying simply on the effect of the pinholes. He submitted many designs, with various arrangements of holes. Some "glasses" used arrays of holes of varying sizes or spacings, allowing viewers to use them for both close-up and distance viewing. As he noted in his patent, he found that holes 0.3 mm in size were good for close-up viewing of minute objects, while 1.5 mm pinholes were good for distance viewing (just about the size of the Xinjiang "spectacle" holes, I note). Guthrie, a doctor himself, suggested that these spectacles be used in cases where a single lens could not correct all the aberrations of the eye.

Guthrie was a fascinating individual, whose contributions to vascular surgery were impressive. It has been argued that he deserved a share of the 1912 Nobel Prize in Physiology that went to Dr. Alex Carrel.[7] He also performed early work on transplant surgery, and, in his most interestingly bizarre experiment, attached the head of a dog to the neck of another dog on May 21, 1908. His work on physiological optics was clearly a foray into a very different line.

These patents were intended for patients with severe visual problems, at a time when exotic asymmetric and aspheric corrective lenses were not yet available, and modern corneal sculpting techniques were not yet developed. For patients with conditions such as keratoconus or irregular corneal shapes, pinhole glasses were often the only real solution. Pinhole glasses were also used for patients recovering from detached retinas. Pinhole glasses were not intended by the medical community to be substitutes for corrective lenses for most patients.

The use of pinhole glasses seems to have started being used as an alternative to regular refractive glasses coincident with the rise of "alternative" medical practices not recognized by orthodox medicine (and often refuted by them), such as the Bates method. The Bates system is a series of exercises for the eye originally propounded by William Horatio Bates (himself a physician) from 1920 to his death in 1931, and continued to this day by organizations. Bates's claim that poor vision is the result of eyestrain, and that accommodation is the result of deforming the eyeball (rather than the crystalline lens). The Bates method does not really suggest the use of pinhole glasses, and actually suggests their use is a hindrance to their method, but many people who use the method also have used pinhole glasses in place of normal refractive lenses. One famous case was writer Aldous Huxley.[8] Daughter-in-law Ellen Hovde recalls that: "He wore pinhole glasses because it concentrated his vision. He did Bates exercises every day." Huxley had extremely poor vision, 20/100, after an attack of keratitis, and wrote *The Art of Seeing* in 1942 about his recovery using the Bates method, although his vision always was very poor. Andy Warhol also reportedly used pinhole glasses for vision correction.

Pinhole glasses start showing up in articles and advertisements from the late 1970s onwards, becoming common in the 1990s in advertisements in tabloid newspapers[9] and now, with the Internet, becoming extremely common online.

These multiple aperture pinhole "glasses" allow you to see over a broader range than a single pinhole, but the multiple images do not "stitch" cleanly together to form a single scene, as ordinary glasses would. Instead, each pinhole gives a slightly shifted view relative to the one next to it, but there is much repetition of the scene, so that one sees a repetitive "mosaic" of images, much like the usual depiction of a "fly's eye" view. This takes some getting used to. But each "scenelet" is in much sharper focus than a myopic person would see without them, so that, by shifting your eyes around, you can take in the entire scene, but only a bit at a time. This is the price you pay for using low-tech pinhole glasses in place of refractive correction.[10–16]

The question is whether such multiple pinhole glasses are really a good long-term solution to eye problems. Are these really good for your eyes, and can you bring the worldwide ophthalmological conspiracy to its knees, or can these potentially harm the wearer? Certainly some proponents of the glasses feel that they are the underdog in this fight. A group calling itself the International Myopia Prevention Association[17] thinks that myopia is caused by eyestrain and exacerbated by eye doctors prescribing glasses, and that pinhole glasses (along with "plus power" reading glasses) are the solution. They cite Guthrie's work on a page about the "Anti-Pinhole Conspiracy," without noting that he felt his pinhole glasses were for extreme cases of aberration. The site *Pinhole-Glasses.com*, on the other hand, advocates pinhole glasses because it feels that minus refractive lenses accelerate myopia. *Ophthalmology Review* looks as if it's a serious eye journal, rather than the pro-pinhole site it really is.[18]

The federal government has raided pinhole glass warehouses and seized such products, but in these cases they have claimed that the vendors made excessive claims for the pinhole glasses being able to cure poor eyesight.[19] Some of the downsides of pinhole glasses are reduced illumination (Wittenberg estimates only about 7 percent of what you see without them, or with standard lenses), loss of peripheral vision, deceptive depth perception, and no blocking of UV light.

There are no articles in refereed journals suggesting that pinhole glasses provide any lasting benefits, and no such articles endorsing their use. Sites where actual doctors are consulted about them uniformly discourage their use.[20, 21] But any Internet search will leave you deluged with pro-pinhole glass websites and positive testimonials—finding the few skeptical sites and real medical advice sites actually takes quite a bit of work and persistence.

The disappointment with pinhole spectacles has a long history, even going back before Heilborn's patent. In his 1875 book *A Practical Treatise on the Diseases of the Eye*, Henry Haynes Walton[22] has this to say:

> *Stenopeic spectacles*, literally, peep-hole spectacles, are better known in science than in practice, because they do not afford all the benefit that their theory holds out, but they have some practical advantage, and besides, afford assistance in examining the optical defects of the eye.
>
> Speaking of the appliance, as for one eye, it consists of a goggle, which so fits the eye as to exclude the light, and in front, the diaphragm has a hole or a slit according to the circumstance of the case. In the test apparatus, there is provision for trying holes and slits of different sizes in succession. The intention is to cut off all the light, and to allow rays to enter only from a certain direction. It is many years since Millard of Oxford Street made several of these stenopeic tests for me, when I was working on the subject of conical cornea.
>
> . . .
>
> It is well known that, by looking through a small aperture, and so diminishing the circles of dispersion, objects can be seen with accuracy much nearer to the eye, that is under a greater angle.... But there are disadvantages in loss of light, and loss of the extent of the field of vision.
>
> Lenses may be combined with the stenopeic spectacles. If such combination be adopted for myopia, in order to avoid loss of light, the slit should be next to the eye. When greater distinctness is required, the lens should be next to the eye.

Walton wrote 20 years before even Heilborn's patent. About the only thing he missed was using multiple pinholes or slits instead of a single one, but his criticisms are as valid today as they were in 1875.

Notes

1. William Hanks Levy, *Blindness and the Blind: A Treatise on Typography* (1872), 504. Online at http://books.google.com/books?id=FTUDAAAAQAAJ&pg=PA504&dq=%22History+of+Spectacles%22&hl=en&sa=X&ei=ChezT__cOYaJ6QGE4-muCQ&ved=0CEAQ6AEwAA#v=onepage&q=%22History%20of%20Spectacles%22&f=false, accessed May 15, 2012.

2. Aristotle, *Generation of Animals and Other Writings*, trans. A. L. Peck (1943), sections 780b15 and 781a5, pp. 503 and 505.

3. "T.W." of St. Cloud, Minnesota in the June 2007 issue of *Backpacker* Magazine ("Reader Tips" on p. 54).

4. Reported in Lea Hyvärinen, *Øjet og Synet* ["Eyes and Vision"] (1991). See also her website at http://www.lea-test.fi/en/eyes/specs.html, accessed May 15, 2012.

5. J. William Rosenthal, M.D., Spectacles and Other Vision Aids: A History and Guide to Collecting (1996), 63. http://books.google.com/books?id=lp4LAAAAIAAJ&pg=PA63&dq=pinhole+spectacles&hl=en&sa=X&ei=Jx2zT-3GFYfA6AHF69GtCQ&ved=0CFAQ6AEwAA#v=onepage&q=pinhole%20spectacles&f=false, accessed May 15, 2012. See also http://www.bowers.org/files/SilkRoad_Teacher_Packet.pdf, p. 8. The relevant item is a set of Eye Shields excavated from Tomb 227 Astana, Turfan, Xinjiang Uygur Autonomous Region Museum Collection. This example measures 15 cm by 5.1 cm. Other examples can be found at http://penn.museum/blog/fun-friday-image-of-the-week/fun-friday-image-of-the-week-sunglasses-from-7th-9th-century-ad/, accessed May 16, 2012; http://ancienthistory.about.com/od/chinasilkroad/ig/Silk-Road-Artifacts/Bronze-Eyeshades.htm, accessed May 16 2012; http://www.antiquespectacles.com/slide_shows/eskimo/eskimo.htm http://www.michelangelo.cn/download/08_Genetics/new_pdf/The_Hungarian_Death_mask.pdf; Chengyuan Ma and Feng Yue Shanghai, *Archaeological Treasures of the Silk Road in Xinjiang Uygur Autonomous Region = Xinjiang Weiwuer zizhiqu silu kaogu zhenpin* (1998), 198; http://news.frbiz.com/women_dressed_in_xinjiang_wrapped-274512.html, accessed May 16, 2012.

6. Charles Claude Guthrie, M.D., Ph.D., "Physiologic Lensless Spectacles," *Archives of Ophthalmology* 11 (2): 254–261 (1934), see http://archopht.jamanetwork.com/article.aspx?volume=11&issue=2&page=254, accessed May 16, 2012.

7. See Hugh E. Stephenson, Jr., and Robert S. Kimpton, *America's First Nobel Prize in Medicine or Physiology: The Story of Guthrie and Carrel* (2001) and http://www.ncbi.nlm.nih.gov/pmc/articles/PMC116749/, accessed May 16, 2012.

8. David King Dunaway, *Aldous Huxley Recollected: An Oral History* (1995), 60.

9. See this example from 1993, http://books.google.com/books?id=Ru4DAAAAMBAJ&pg=PA65&dq=%22pinhole+glasses%22&hl=en&sa=X&ei=Arq1T4W5JZS36QGN7tnNCg&ved=0CF0Q6AEwATg8#v=onepage&q=%22pinhole%20glasses%22&f=false, accessed May 17, 2012.

10. S. Wittenberg, "Pinhole Eyewear Systems: A Special Report," *J. American Optom. Assoc.* 64 (2): 112–116 (February 1993).

11. Christopher Jargodzki and Frank Potter, *Mad about Physics: Braintwisters, Paradoxes and Curiosities* (2001), 170.

12. M. P. Keating, "Reading Through Pinholes: A Closer Look," *Am. J. Physics* 47 (10): 889–891 (October 1979).

13. S. S. Mathur and R. D. Bahuguna, "Reading with the Relaxed Eye," *Am. J. Physics* 45 (11): 1097–1098 (November 1977).

14. Giuseppe Colicchia, Martin Hopf, Hartmut Wiesner, and Dean Zollman, "Pinhole Glasses," *The Physics Teacher* 46 (1): 26–28 (January 2008).

15. R. Petera, "Pinhole Glasses," *The Physics Teacher* ***16*** (6): 383 (September 1978) and 44 (2): 122 (February 2006).

16. Mojca Čepič, Ana G. Blagotinšek, and Nada Razpet, "Looking Through Pinhole Glasses with a Digital Camera," *The Physics Teacher* 46 (3): 186 (March 2008).

17. http://www.myopia.org/conspiracy.htm, accessed May 17, 2012.

18. http://www.noorvision.com/weblog/archives/2003/07/, accessed May 17, 2012.

19. http://www.quackwatch.org/01QuackeryRelatedTopics/eyequack.html, accessed May 17, 2012.

20. http://ehealthforum.com/health/pinhole-glasses-t270363.html, accessed May 19, 2012.

21. http://www.medhelp.org/posts/Eye-Care/Pinhole-Glasses-for-High-Myopia/show/1735555, accessed May 19, 2012.

22. Henry Haynes Walton, A Practical Treatise on the Diseases of the Eye (1875). http://books.google.com/books?id=mf40Q94KC20C&pg=PA256&dq=%22stenopeic%22&hl=en&sa=X&ei=XTW3T86cDoj1ggfq27XACg&ved=0CFQQ6AEwBA#v=onepage&q=%22stenopeic%22&f=false, accessed May 19, 2012.

{ 33 }

Undulations

In writing the essays which eventually appeared in *The Spectrograph*, the house organ for M.I.T.'s George Harrison Spectroscopy Lab, I tried to find topics that were somehow related to spectroscopy, for obvious reasons. And, since I have always been interested in origins and beginnings, I looked to where ideas and practices have their start. One often finds unexpected and interesting things there. Several such lines of inquiry led me to the same question—who first measured wavelength?

Curiously, I can find no history of wavelength. (The best general reference is the work of Hearnshaw, cited below.) There are no articles or books devoted to it, and the subject is neglected in the classic optics texts. You would think that the originators of the Wave Nature of Light would have addressed the question of wavelength as an essential part of the theory. Yet neither Christiaan Huygens nor Robert Hooke say anything about it.

Looking from the other end, one widely used unit for the wavelength is the *angstrom*, named after Anders Jonas Ångström of Sweden, who from 1861 to 1868 measured a total of 1,000 lines of the solar spectrum to such accuracy that it was the standard for many years. To do this, he used a 2,000 line/cm diffraction grating. But Ångström was not the first to measure solar spectra, and he acknowledged his debt to Joseph Fraunhofer, who in the 1820s had observed 574 lines of the solar spectrum and gave careful measurements of about 350 of these.[1]

Fraunhofer made his own grating for performing his measurements by winding fine wire on a frame made of screw threads. His first wire grating had 260 wires, which he improved upon with a 325 wire/inch grating. He later constructed a ruling engine to produce diamond-ruled grooves on a glass substrate. In 1822 he produced one with 3,340 grooves per inch, and increased that to 7,790 grooves/inch in 1823.[2] But Fraunhofer did not claim to be the first to measure wavelength. His predecessor was Thomas Young, that indefatigable champion of the wave nature of light Young looked everywhere for proofs that light was, in fact, a wave phenomenon. He is probably most famous for "Young's two-slit experiment," where light from a distant source is directed through a pair of slits, and the region where illumination from both slits overlap exhibits interference between the two slits, creating a setoff evenly spaced alternating light and dark bars. (In fact, there's no evidence that Young performed this experiment as it's normally described, and it may just have been a *gedankenexperiment*.) But he went well beyond this, seeing evidence of such interference in the *supernumerary bands* of rainbows (which no one had previously explained satisfactorily), in the interference from scratches on surfaces, and in what seems to be the first diffraction grating experiments.

He described these in the second Bakerian Lecture he gave, in 1802.[3] Having first described how the "two slit" experiment would work with point sources, he goes on to say that "in order that the effect may be the more perceptible, a number of points must be united into two parallel lines; and, if several such pairs of lines be placed near each other, they will facilitate the observation."

The actual devices used for his experiments were very interesting—he didn't construct them himself, or have someone else build custom apparatus for him—he used what we would today call "commercial off the shelf" (COTS) parts. "The best subjects for the experiment are Mr. Coventry's exquisite micrometers; such of them as consist of two or more parallel lines drawn on glass, at the distance of one five hundredth of an inch, are most convenient. Each of these lines appears under a microscope to consist of two or more finer lines, exactly parallel, and at the distance of somewhat more than a twentieth of that of the adjacent lines."

The Mr. Coventry referred to was a famous instrument manufacturer who lived in Southwark and died before 1816. He is acknowledged in other papers of the time, and as a manufacturer of crosshairs and reticles for telescopes and microscopes.[4] These were called "micrometers" at the time because of their use in measuring small distances.[5]

In this case, Young certainly *did* perform the experiment and found that he saw bright red beams—indicating that the path length differences between light reflecting from the lines were integral numbers of wavelengths—at angles that were integral multiples of the sines. On the basis of his measurements, Young was able to provide a table of measurements of wavelengths associated with Newton's seven colors (table 33.1).[6]

But Young did not work in a complete vacuum. Even before he had measured the wavelengths, he had an idea of what their relationships ought to be: "the undulations of red, yellow, and blue, to be related in magnitude as the numbers 8, 7, and 6" and, indeed, from his own numbers red:yellow is about 1.13, while 8:7 is 1.14, and red:blue is 1.31, while 8:6 is 1.33.

Where did Young get these numbers? From Sir Isaac Newton. In his own Bakerian lecture, Young quotes at length from Newton's *Opticks* and from his articles in the *Philosophical Transactions*. As the history of optics is now commonly taught, Newton was the great champion of the Corpuscular Theory of Light, in which light is composed of a plenitude of tiny particles, in distinction to the wave theory of Huygens, Hooke, and Young. But it is worth pointing out that Newton

TABLE 33.1 Measurements of Wavelengths Associated with Newton's Seven Colors

Color	Wavelength (inches)	Wavelength (microns)
Extreme Red	0.0000266	676
Red	0.0000256	650
Intermediate	0.0000246	625
Orange	0.0000240	610
Intermediate	0.0000235	597
Yellow	0.0000227	577
Intermediate	0.0000219	556
Green	0.0000211	536
Intermediate	0.0000203	516
Blue	0.000196	498
Intermediate	0.0000189	480
Indigo	0.0000185	470
Intermediate	0.0000181	460
Violet	0.0000174	442
Extreme Violet	0.0000167	424

had first performed many experiments that suggested that light might be a wave, and that, moreover, the lengths of the wavelengths appeared to be correlated with the colors observed.[7]

> *Hypothesis III: The Sensation of different Colours depends on the different frequency of Vibrations, excited by Light in the Retina....* the agitated parts of the bodies, according to their several sizes, figures, and motions, do excite vibrations in the ether of various depths or bignesses, which, being promiscuously propagated through that medium to our eyes, effect in us a sensation of light of a white colour; but if by any means those of unequal bignesses be separated from one another, the largest beget a sensation of a red colour, the least or shortest of a deep violet, and the intermediate one of intermediate colours.

Newton's observation and experiments are described in detail by Newton himself in the Second Book of his *Opticks*, Parts 1 and 2. He first noted it by pressing the faces of two prisms together. Not being quite flat, and slightly convex, they rocked against each other about the point where they touched. The point of contact was clear, but in concentric irregular rings around this point he saw bands of color.

More careful measurements made with spherical parts of known large radius of curvature, abutting plane pieces of glass, or other large radius spheres, produced much more regular such color "haloes," circular in shape and regular in diameter. Knowing the geometry, Newton was able to calculate the surface sag and associate the sag and its integral multiples with particular colors, and from this he correlated the colors with characteristic lengths. Drawing an analogy with the notes of a stretched string, he could model the colors as vibrations whose lengths would probably exist in the same sort of ratios as musical notes (and impelled him to see seven colors, as there were seven notes in an octave, not counting the highest one (the overtone)). He even observed that wave effects might cause the observed colors in regular media, such as feathers (see chapter 7, "Roy G. Biv").

Newton became dissatisfied with the wave theory, however, despite these successes, citing, among other things, the lack of light diffracted into shadow. This, despite the fact that he cites the experiments of Francesco Maria Grimaldi, which seem to show wave behavior. Newton's objection is the same that Poisson would later make, and it's sobering to think that if he had performed Arago's experiment and seen that on-axis spot, we might have had the Wave Theory of light established a century earlier.[8]

So the answer to "Who First Measured the Wavelengths of Light and Associated them with Colors?" is really Isaac Newton—although he may have changed his mind about their significance. As for how he measured the wavelengths, he used the small but calculable sag of a spherical surface to do so. David Rittenhouse came extremely close to being the first to measure the wavelengths of light with his first-ever diffraction gratings, but the press of time prevented his following up his experiments. Thomas Young was the next to investigate the wavelengths of light and seems to be the first to really state and believe the model, and to provide explicit wavelengths for a variety of colors. His measurement used COTS telescope reticles, pressed into service as diffraction gratings. Fraunhofer and others followed in their footsteps, constructing better and finer instruments and performing more detailed and precise measurements.

Notes

1. J. B. Hearnshaw, *The Analysis of Starlight*: One Hundred and Fifty Years of Astronomical Spectroscopy (1986), 27.

2. George R. Harrison, "The Controlled Ruling of Diffraction Gratings," *Proc. Amer. Phil. Soc.* 102 (5): 483–491 (October 20, 1958). He, in turn, cites J. Fraunhofer, "Kürzer Bericht von der Resultaten Neuere Versuch uber die Gesetze des Lichtes und die Theorie Derselben," *Ann. Der Physik* 74: 337–378 (1823); see also John Hearnshaw, *Astronomical Spectrographs and their History* (2009), 2.

3. Which was reprinted in the *Philosophical Transactions of the Royal Society of London* 92: 12–48 (1802).

4. Besides this paper of Young's, he is referred to in Ez. Walker, "On Micrometer Telescopes," *The Philosophical Magazine*, Series 1, 47 (213): 14–16 (1816); David Brewster's book *A Treatise on New Philosophical Instruments* (1813), Book 1, p. 66; and again by Young in "On Changeable Colours and Glories," *Philosophical Magazine*, Series 1, 42 (186): 292–296 (October 1813).

5. The first use of a crosshair or reticle placed at the intermediary image of a Keplerian telescope is attributed to William Gascoigne (1612–1644), an amateur telescope maker who

found that a spider had built its web inside his telescope tube, and that the web was in perfect focus when it passed through the common foci of the lenses. This, he realized, could be used to make a system for performing accurate measurements and kept this technique a secret. See chapter 5, "The Miracle of Saint Gascoigne."

6. There are two earlier claims of diffraction gratings. David Rittenhouse of Philadelphia, a noted astronomer who measured the Transit of Venus, director of the Mint, and official in charge of the Mason-Dixon Line. In 1785 a question was posed to him by Francis Hopkinson, one of the signers of the Declaration of Independence. He had held a silk handkerchief stretched out in front of him and looked at a distant street light, and was surprised that the cross-hatch pattern he observed did not move when he moved the handkerchief. Wasn't this a magnified image of the threads in the handkerchief? He presented the problem to Rittenhouse, who saw in it an example of the diffraction that Newton had referred to. To study it more closely, he constructed a "square of parallel hairs about half an inch each way." He got a watchmaker to cut very fine threads onto brass wire, 106 to the inch, and carefully stretched 50 to 60 hairs across these. Looking at a distant source through these, he saw three parallel lines of light. He replaced the hairs by thicker ones, 190 to the inch and about 1/250 of an inch apart. With this he could see six lines—six orders, in modern terms—on either side, with bluish coloration on the inner side and reddish on the outside—the opposite of a rainbow or prismatic spectrum. He noted that the angles of each order formed integral multiples of the first order, but did not take his observation further and determine the wavelengths. "By pursuing these experiments," he wrote, "it is probable that new and interesting discoveries may be made respecting the properties of this wonderful substance, light.... But want of leisure obliges me to quit the subject for the present." As far as is known, he never took it up again.

The details appear in "An Optical Problem," *Trans. American Philosophical Society* 2: 201–206 (1786). It is reprinted in its entirety, with comments, in T. D. Cope, "The Rittenhouse Diffraction Grating," *J. Franklin Institute* 214: 99–104 (1932). See also chapter 9, "Hopkinson's Silk Handkerchief."

The other prior claim to diffraction gratings is that James Gregory discovered the principle of the grating in the seventeenth century, but his observations were on finely grooved items such as bird feathers, and he doesn't appear to have created any gratings himself, or to have performed measurements with them.

7. *Phil. Trans.* 7: 5088 (1672).

8. See also chapter 7, "Roy G. Biv." Sometimes Newton's refusal to see the evidence for the Wave Nature of Light seems perverse.

PART III

Pop Culture

{34}

This Is Your Cat on Lasers

There are several examples that show up when people write about patents that they can't believe were ever granted. One of the most prominent of these is Patent No. 5,443,036, granted August 22, 1995, for a *Method of Exercising a Cat.*[1] This patent essentially states that you can use a hand-held laser to cast a spot of light on an opaque surface so that your cat will chase it. The patent does not explain or cover the device itself, nor is the laser built into some mechanism that redirects it (there are several such laser-toy patents). This patent is merely for the idea of using a laser to play with your cat.

The reaction of most people is "They got a patent for *that*?! I do that all the time!" Indeed, playing with cats (and dogs, and probably other inquisitive animals) with easily available hand-held lasers is incredibly common. The idea is by no means new. I had several friends who endured the effort and fatigue of holding a power cord-connected helium neon laser head and playing with their cats as long ago as 1980. I feel certain that people had been doing the same thing much earlier, perhaps using mirrors to reflect light from older, heftier lasers that weren't easy to lift and maneuver. Certainly there are plenty of vocal people claiming that they were doing this before the patent was granted in 1995.

Others object to the idea that so simple and obvious a concept as Using a Laser to Play with a Cat can even *be* patented. It seems to fail the test of being something new and original and nonobvious to those *Familiar with the Art*. If, as the classic example goes, you can't get a patent for putting an eraser on the end of a pencil, why should you be able to get a patent for this? A lot of humor sites list it among "Stupid Patents."[2-5] Other websites see the granting of this patent as a sign of the decline or failure of the Patent System itself.[6,7]

I got the impression that most of these people have not actually seen or read the patent, and that it was high time that someone did so. My first step was to try to contact the inventors and patent holders. I tried via telephone, mail messages, and e-mail, but with no success. In the absence of their personal information, I tried to glean what I could from the patent itself, as well as the supporting documentation that is publicly available on the patent office website and from the patent office itself.

The first thing you notice, after the title, is the list of references. These include two articles on cat physiology. There is also a list of patents that refer to light-related toys and to laser pointers.

On the next page are the two figures illustrating the patent's contents. Patent drawings range in style from excellent mechanical drawings to crude depictions. These fall at the "crude" end of the scale, a sort of techno-naïve art.

The description is straightforward if stodgy, in the usual literary style of patents. Finally, the patent states the heart and soul of the document, the Claims. These state that this is a method of giving a cat aerobic exercise by using a laser beam, directed by hand onto an opaque surface and moving the spot around. Reading this, I began to suspect that the patentees were not entirely serious.

The patent was originally submitted November 2, 1993, and rejected on August 17, 1994. The reason given for the rejection was that the idea being patented was not original. The examiner cites the work of R. Carayan, E. Mocaër, and M. Fabre-Thorpe.[8] The researchers had trained or selected cats that would respond to the appearance of a spot of light. They harnessed these cats into position and placed them in a darkened box, then observed how quickly they responded to the appearance of the light after a drug had been administered. The examiner also cited US Patent No. 3,877,171. This was for a "Flashlight Amusement Device," granted in 1975 to Conrad B. Sloop and Wanda L. Clearwaters, and assigned to Mattel toys.. It basically consists of a cap that can be attached to a standard flashlight, containing a mirror that redirects the light by ninety degrees, and can rotate. It's intended to amuse children. Commenting on his "Prior Art," the examiner noted that "it would have been an obvious to one skilled in the art manually manipulate the laser apparatus in the method disclosed by Carayan et al. in order to produce a more irregular and erratic movement thereof" [*sic*].

This was termed a "Non-Final Rejection," and in such cases the prospective patentee typically responds by modifying his claims to make the patent acceptable. They did so on November 1, 1994, adding the italicized words to Claim 1:

A method of exercising a cat *under normal lighting conditions*...

In other words, they apparently figured that the cat in the French experiment was in a dark box. Well, their laser could be used in normal lighting!

The Patent Examiner's reply came two months later, on January 9, 1995—Carayan et al. hadn't said what lighting conditions they used. Therefore, whatever they used were "normal lighting conditions." Application refused. In fact, it was made a Final Rejection.

Normally, that would have ended the patent quest. But the Patent Office allows a Request for Continued Examination (RCE) upon payment of a fee.[9] They scheduled a telephone interview with the examiner and sent a copy of a videotape showing "implementation of the applicants' invention"—presumably one of them playing with a cat with a laser pointer. They pointed out that their claim of using a laser differentiated it from prior claims about the use of flashlights. And they amended their all-important claim 1 to read:

> A method of *inducing aerobic exercise in an unrestrained cat*

Evidently this distinguished it from the French experimental work since, in that work the cat *was* restrained.

You might think these changes were minor, cosmetic, and would not make any substantive difference, but the Patent Examiner allowed the changes after this.[10] On August 22, 1995, the patent was issued.

Did the patent do the patentees any good? I have been unable to locate any evidence that this was licensed out. None of the laser toys reference it, and there does not seem to be a company associated with the patent holders. Manufacturers and distributors of pet lasers that I contacted tell me that they were not contacted by the patent holders, and did not seek them out. I did locate one individual who claimed to have written to the patent holders. He wrote them, asking to license their patent:

> This is regarding your patent, **US05443036.** I would like to talk to you to negotiate an agreement wherein I could exercise my cat with a laser pointer. I would be willing to pay a reasonable flat rate for a one-year license. Please let me know what your conditions are, and I'm sure we can work something out.

The response he reported getting was terse: "*That standard fee is $20,000. Thank you.*" You won't be surprised to learn that he did not pursue this.

Today you can walk into just about any pet store and have you choice of several different hand-held laser pointers that are sold explicitly as pet toys—usually as cat toys. How can this be? Isn't the patent still in force? Patents granted since June 1995 should have a life of 20 years since the earliest filing date, during which time the holders can exclude others from using their work, unless they have secured permission, usually by paying fees. Such "cat lasers" appeared long before 2012. How can this be?

Since 1980, the Patent Office has required the payment of scheduled fees for maintenance. This is similar to the annual patent fees charged by some countries. In the United States, there are three fees, due 3½ years, 7½ years, and 11½ years from the granting of the patent, and steadily escalating in cost. The Patent Office evidently counts on the patents to be a financial success. As of this writing, the fees for a Small Entity (as opposed to a large company) are $465 for 3.5 years, $1,180 for the 7.5 year patent, and $1,955 for the 11.5 year fee. I suspect that, whatever the intent in getting the original patent, the cost of maintaining it with little or no licensing fee was getting too much to keep up. In the absence of the 11.5 year maintenance fee, the patent expired on September 17, 2007. After that time, anyone could manufacture the item for the purpose of exercising a cat and not have to pay a fee.

So, after all that, why was the patent issued, and why couldn't it have been stopped? Should it have been stopped? Several Internet sites note that the 1982 book *187 Ways to Amuse a Bored Cat* describes the same basic idea, but using a flashlight instead of a laser.[11]

But that's not really correct. The book, edited by Howe C. Stidger and Ruth Stidger, includes as entry No. 70 a suggestion by Kathy Shaw of Deer Park, Washington, to shine a spot from a flashlight in a darkened room for cats to chase. But as we've seen, the Patent Examiner was already presented with the concept of a hand-held flashlight pointer in the form of US Patent No. 3877171. In addition, the darkened room specification would have been struck down by the first revision of the patent, and at their second hearing the patentees argued that a laser was intrinsically different from a flashlight.

Those familiar with the workings of the patent office suggest that patents like this simply aren't worth the trouble of fighting, as they are not enforceable, and will fail at the first serious challenge. The implication is that Patent Examiners are overworked, and it's simply not worth their while to put so much effort into fighting a patent that won't really stand up. "Bad patents, such use of a laser wand to exercise a cat, aren't going to have much impact in terms of litigation; in terms of the time lost, messing around with toothless applications doesn't help speed the legitimate patents through the process," writes Mark Lemley, law professor at Stanford University.[12] Reviewing a book on the patent office, Thomas G. Field, law professor at Franklin Pierce Law Center, wrote:

> Three [patents] mentioned in their book and elsewhere are for sideways swinging, **exercising cats with laser pointers**, and wristwatches for dogs. They might not resort to such examples, however, if they appreciated that the first is unenforceable, **the second could not be enforced except possibly under § 271(b),** and the third if enforced would seem unlikely to endanger much of anything.[13]

Regarding the cat laser patent, one of the manufacturers I talked to said that he wasn't concerned. I noted that his cat laser toy had a notice of "Patent Pending" on it, which seemed unusual, under the circumstances. The reason it was there, I was told, was that it was for a Design Patent, covering the actual design of the device and preventing anyone else from using their prints and tooling from using them to manufacture their own laser pointer cat toys.

As for that other case mentioned by Professor Field, the *Method of Swinging on a Swing*, that's US Patent No. 6,368,227, granted April 9, 2002. I'll let *Playground Magazine* inquire into that one.[14, 15]

Notes

All websites accessed August 7, 2008, unless otherwise indicated.

1. Martin Gardnet *Fads and Fallacies in the Name of Science* (1957), 347. Scholfield's patent was actually cited by another—Bernt Schroer's 2003 patent *Torsion spring for counter balancing weights*, Patent No. 6,527,264. I suspect Mr. Schroer wasn't being entirely serious.

2. http://obsidianwings.blogs.com/obsidian_wings/2003/12/stupid_patent_t.html.

3. http://www.freepatentsonline.com/crazy.html.

4. http://blog.simplyjean.com/2008/05/27/imaging-linking-is-patented-websites-and-blogs-affected-advertlets-nuffnang-and-blog2u-may-have-to-call-it-a-day/.

5. http://www.maximumpc.com/article/daily_news_brief_ultra_goes_on_suing_spree_over_modular_cables.

6. http://www.eff.org/deeplinks/2008/06/patent-office-cant-do-it-all-alone.

7. Mark Lemley, Doug Lichtman, and Bhaven Sampat, "What to do about Bad Patents?" http://www.law.uchicago.edu/files/lichtman/bad-patents.pdf. A more serious discussion is given in Christopher L. Culp, *Structured Finance and Insurance: The Art of Managing Capital and Risk* (2006). See also Anu R. Sawkar, "Are Storylines Patentable? Testing the Boundaries of Patentable Subject Matter," *Fordham Law Review* ***76*** (6): Article 13 (2008), which discusses the patent and its limits. Available online at http://ir.lawnet.fordham.edu/cgi/viewcontent.cgi?article=4371&context=flr, accessed July 25, 2012.

8. *Psychopharmacology* 104 (3): 328–336 (1991).

9. On the patent examination procedure, see http://pagebox.net/exam1ForPrinters.html.

10. http://www.bedope.com/stories/0069.html.

11. One location is http://www.freepatentsonline.com/crazy.html.

12. http://blogs.zdnet.com/BTL/?m=20070226.

13. http://www.patentlyo.com/patent/book_review/index.html.

14. http://www.todaysplayground.com/.

15. Since this article was originally published, a Wikipedia page has appeared devoted to it: http://en.wikipedia.org/wiki/Method_of_exercising_a_cat. There has also appeared a new interpretation of the patent art, drawn for an article in the *IEEE Spectrum* by Josh McKible in 2009 that is of much better quality than the actual patent drawing http://www.thelittlechimpsociety.com/2009/02/method-of-exercising-a-cat/, accessed July 25, 2012. Also here: http://www.imagekind.com/Method-of-Exercising-a-Cat-art?IMID=40535733-1893-4de2-8641-77dca13e95a0. I have, by the way, attempted to contact the authors of the patent via telephone and e-mail, but without success.

{35}

Dord

According to *The Straight Dope*, the no-holds-barred answer column of the alternative newspaper *The Chicago Reader*, the likelihood of an etymology being true is inversely proportional to its cuteness. Thus, the story that *Kangaroo* actually means "I don't understand" (the Australian aborigine's reply to a member of Captain Cook's expedition, who tried to ask the name of the animal) is highly doubtful. A similar story is told about the naming of the *Yucatan* peninsula and is equally dubious. As another example, it's said that *Cape Nome, Alaska*, and the nearby town of Nome got their name when a cartographer wrote a notation on the map "? Name," which got misinterpreted as "C. Nome." Clearly, if true, the cartographer had handwriting as bad as mine. It's not likely that the story is true, however.[1]

So it is always fascinating when such a story actually does turn out to be true, and demonstrably so. For a period of about five years, the *Webster's New International Dictionary* (2d ed.) carried the word *dord* between *Dorcopsis* and doré.[2] The story is told in many places, but they all ultimately derive from "The History of Dord," written by Philip B. Gove, the editor of the dictionary himself, and published in the journal *American Speech* in 1954.[3]

According to Gove, it all started on July 31, 1931, when Dr. Austin M. Patterson, special editor for chemistry, sent a 3′′ × 5′′ card to the offices of G. and C. Merriam Company. It was intended as an update to the first edition of the *Webster's New International Dictionary*'s (1909) entry on abbreviations written as "**D.**" On the card was typed "**D or d, cont. density.**" "**D or d**" indicated that the abbreviation was written as the capital or small letter *D* or *d*, without a period, "cont." meant that this slip would be one of several entered in the definition (it stood for "continued"), and "density" was, of course, the definition.

The typist who copied Patterson's "slip" was supposed to put a wavy line under "**D**" and "**d,**" indicating that they were to be set in boldface type. Instead, the typist put a continuous wavy line under the entire phrase "**D or d,**" also added a label "**Physics and Chem.**" and marked the first letter of "density" as a capital. From then on it was treated as a word in its own right.

The next typist repeated "Dord" with a wavy line under it all, typed "**Cont**." with an initial capital letter (it was afterwards crossed out altogether). It started to achieve a life of its own when someone added "n.," indicating that this new word was a noun, something it clearly was not. Its life was put beyond question when someone fabricated a pronunciation for it—"**dôrd.**" The word had no etymology, which should have aroused the curiosity of the etymologist, but didn't. It passed the proofreader's scrutiny without any problem and was duly included in the next edition.

Dord lived a short and happy life as a "ghost word"[4] for five years until an office editor noticed on February 28, 1939, that the word had no etymology and submitted a slip reading "**plate change / imperative / urgent.**" This being the era before computer setup of printing plates, the problem was erased by physically cutting the offending line from the plate, shifting some lines up, and expanding the definition of *doré furnace* from "a furnace for refining doré bullion" to the slightly longer "a furnace in which doré bullion is refined," thus eating up the excess line. Dictionaries without *dord* began to appear in 1940.

A lot of work, and Gove concluded wistfully: "Probably too bad, for why shouldn't *dord* mean "density"?

An interesting story, told and confirmed by the editor himself. Nevertheless, after I heard it, I began to have my doubts. My first thought, optical engineer that I am, was that although I've seen **D** as an abbreviation for *optical density* often enough, I've never seen **d** used that way. In fact, you would think that **OD** would be an alternate designation. It didn't strike me until later that they probably meant mass density, not optical density. But then, in my experience, the symbol for mass density is usually **ρ** (it's also the cgs symbol for charge density in electrostatics). But perhaps that's a relatively recent innovation.

I did an informal survey of the Internet and old physics texts. Both **D** and **d** are used for mass density, with the former outnumbering the latter about 10 to 1. Rho (**ρ**) is used, too, of course. Optical density is represented by **D** usually, but never by **d**. Very often today it is abbreviated **OD**, so as not to be confused with other abbreviations using the single letter **D**. In old papers about photographic density and the Hurter-Driffield curve, the optical density is represented not by **D**, as it is today, but by **Δ**.

I decided to check dictionaries to see what they had to say today. None of them has *dord*, I find, even as an eccentricity. But, to my surprise, it wouldn't even be possible to make the same mistake today as had been made 70 years ago—no modern dictionary lists both **D** and **d** as abbreviations for *density*! *The Oxford English Dictionary* (2d ed.), *The American Heritage Dictionary*, and *Chambers's 21st Century Dictionary* don't list either **D** or **d** as abbreviations for *density*. The *Random House Dictionary* lists both **D.** and **d.** as allowed abbreviations. But note the very significant periods following them (it lists other definitions for **D** and **d** *without* periods, but *density* is not among them). The *Encarta World English Dictionary* (1999) gives **d** as an abbreviation for *density*, but not **D**, **D.**, or **d.** Curiously, it gives **D.** (with the period) as an abbreviation for *Diopter*, which I have never seen, but ignores the usual **D** for *Diopter*. *The Oxford Dictionary of Abbreviations* (2d ed., 1990) lists **d** as an abbreviation for *relative density* and **d.** (with a period) for *density*.

The *Webster's New Universal Unabridged Dictionary*, Deluxe Second Edition (1983), arguably the descendant of the one that gave us *dord* in the first place, lists only **D** as an abbreviation for *density*. But, as if asking for another opportunity to let the word spring into existence, the definition above for **D, d** has, as its first definition "**of D or d,**" as if tempting fate. All that would be needed is to close those gaps. . . .

All of this makes me realize how extremely unlikely that original coinage of *dord* really was. If the word had been written in practically any other way it would have excited enough interest to snuff out its fledgling life. The presence of a Greek letter, like **Δ** or **ρ** (*Δord*), or of a period after either symbol (*D.ord.*), or if **OD** had been used, turning it into a two syllable word with excess capitals (*DorOD*, or *ODord*), would have given away the game. Even had the capital and small letter forms been reversed, it would have attracted attention (*dorD*).

It's probable that *Dord* was just very lucky that none of these possibilities arose, and that the "word" formed by chance had both an obvious form (noun, based on the definition) and an obvious pronunciation (*dôrd*, since all other English words ending with *–ord*, such as *lord* or *ford*, are pronounced that way). Had any of these chance possibilities been violated, *Dord* would have been exposed as the fraud it was before making it into the published dictionary. There exists a small possibility in my mind that the word was intentional, a prank that perhaps got out of hand. But it seems unlikely.

What I *am* certain about is that dictionaries, despite all the checking and re-checking, still need to be inspected more closely. I've seen **D.** and **d.** (with periods) as abbreviations for *density*. Or **D.** with a period as abbreviation for *Diopter*. Neither **D** nor **d** is listed as abbreviation for *density* in three dictionaries. It's not as bad as the appearance of the "ghost word" *dord*, but these instances don't really reflect current usage.

Notes

1. The Straight Dope column on *Kangaroo, Yucatan,* and *Nome* is online at http://www.straightdope.com/classics/a2_236.html, accessed May 21, 2012.

2. Look them up, lazybones.

3. On "Dord," the original reference is "The History of Dord" by P.B. Gove of Merriam and Company, *American Speech* 29: 136–138 (1954); the urban legend website Snopes covers it at http://www.snopes.com/language/mistakes/dord.htm, which is where I first learned of it. You can also find it at www.fun-with-words.com/websters_dord.html, accessed May 21, 2012. Neither of these reprints the entire Gove article. There is also an account of the affair in Richard Shenkman and Kurt Reiger, *One Night Stands with American History: Odd, Amusing, and Little-Known Incidents* (1980), 232; and in Tom Burnam, *More Misinformation* (1981), 59. Bruce Manning Metzger used the case as an example of textual variation and corruption in

his *The Text of the New Testament: Its Transmission, Corruption, and Restoration* (1968), 207. A later account of it for the same purpose is in *The Bible in its Literary Milieu: Contemporary Essays*, ed. Vincent L. Tollers and John R. Maier (1979), 236. http://books.google.com/books?id=vk8QAQAAIAAJ&q=%22dord%22&dq=%22dord%22&hl=en&sa=X&ei=Wf26T_6KI4Wm6gGzh5DbCg&ved=0CE4Q6AEwBDiMAQ, accessed May 21, 2012. There is an early account of it, pre-dating Gove, in *Coronet* 24 (4): 15 (1948), http://books.google.com/books?id=WK5UAAAAYAAJ&q=%22dord%22&dq=%22dord%22&hl=en&sa=X&ei=9vu6T8DhOMLo6AH6m9X5Cg&ved=0CDsQ6AEwADhu and in *The Saturday Review* 25: 33 (1942), http://books.google.com/books?id=MXw5AQAAIAAJ&q=%22dord%22&dq=%22dord%22&hl=en&sa=X&ei=yPy6T4X2E8bw6AHd-NDRCg&ved=0CDsQ6AEwADiCAQ, accessed May 21, 2012. It also appears in William Morris and Mary Morris, *Morris Dictionary of Word and Phrase Origins*, 2d ed. ([1977], 1988), 188; and in *The British Medical Journal* 316 (7124–7136): 534 (1998).

4. In case you're wondering about other "ghost words," consider "brean," which appeared in the works of Robert Louis Stevenson (a misprinting of "ocean"), "pornial" (a misprinting of "primal"), "Zzxjoanw," supposedly a Maori word for "drum" or "fife" (but wholly imaginary), and "kelmenopy," a word frankly invented by John Ciardi from the names of the letters K, L, M, N, O, and P. These have all appeared in dictionaries (Ciardi's stuck his in his own *Browser's Dictionary*). See www.wordways.com/ghost.htm and www.languagechat.com/archives/000661.php, accessed May 21, 2012. "Dord" is also an English name and the name of an Irish Horn.

{36}

Zap!

The word "ray" comes from the same Latin root as "radius" and "radiate," and implies something flowing outward from a central source. Similar words appear in the Romance languages, and the *Oxford English Dictionary* (*OED*) notes that "ray" was used from the fourteenth century forward. The word didn't become common until the seventeenth century.

Early uses invariably refer to light rays. According to the *OED*, the first time the word was used to refer to the emanation of something other than light was in a description of "magnetick rayes" that dates from 1664. In 1814, nonvisible electromagnetic rays were first mentioned as "rays transmitted from the sun that do not illuminate."

But what really started the ball rolling was the work of Julius Plücker, Professor of Mathematics and Physics at the University of Bonn. He explored the phenomena of gas discharges in nearly evacuated tubes subjected to electrical stimulation. His tubes were made by master glassblower Johann Heinrich Wilhelm Geissler, and they became known as "Geissler Tubes"—a name that has stuck over the years. Plücker found that gas discharge seemed to proceed from particular points on his cathode. If he reduced the cathode to a point, the output seemed to radiate from this point, similar to light beams.[1,2]

Plücker thus became the first to call this new phenomenon "rays," or *strahlen*. Johann Wilhelm Hittorf, working at the Royal Academy of Münster in Prussia observed in 1869 that *kathodestrahlen* could cast shadows if an object were interposed between the cathode and the fluorescing screen, indicating rectilinear propagation in straight lines. He called these "glow rays" (*glimmstrahlen*). It seemed that new and exotic rays were cropping up everywhere.

Study of the new rays proceeded in many places. William Crookes—an independent researcher—found that they could be deflected by magnetic and electric fields, implying that the beam consisted of charged particles. In 1886, Eugen Goldstein of the University of Berlin bored a hole in the cathode of a Crookes tube and found a new ray streaming in the opposite direction. Because the rays went through the channel, they were called *kanalstrahlen*, or "channel rays." The opposite direction of streaming implied that this was a beam of particles having the opposite charge of the cathode rays.

On November 8, 1895, Wilhelm Roentgen at the University of Würtzburg covered his Crookes tube in black paper to block light from entering. He then placed a screen of fluorescent barium plantinocyandide beyond the end of the paper-wrapped tube. When he connected the leads of the induction coil, starting a discharge, the screen nevertheless glowed. This indicated that invisible rays had penetrated the glass walls and the paper. After six more weeks of work, Roentgen presented a paper at the December meeting of the Würzburg Physico-Medical Society, announcing a new and invisible form of radiation called X-rays.

Within months, natural radiation was discovered by Antoine Henri Becquerel of the École Polytechnique in Paris. He acted on the ideas of Roentgen and Henri Poincaré, who suggested that fluorescent materials might be particularly good for producing X-rays. Three years later, Rutherford noted that the radiation had at least two components, which he called alpha rays and beta rays. The following year, Paul Villard of the École Normal of Paris noted the existence of gamma rays—a third, far more penetrating component that could not be deflected by electromagnetic fields.

Around 1914, V. F. Hess and W. Kohlhorster of Germany discovered another type of ray. They were conducting experiments with balloons to observe how the ionization of the atmosphere varied with altitude. After ionization decreased initially with increasing height, it then began to increase rapidly with altitude. The researchers postulated an extraterrestrial origin for the phenomenon, and called the rays *hohenstrahlung* or *ultrastrahlung*. Robert Millikan coined the equivalent term in English in 1925—"cosmic rays."

By the early twentieth century, many so-called rays came to be recognized as what we now call "particle beams." Alpha rays are helium nuclei, and beta rays are electrons. Of the three forms of Becquerel's radioactivity, only gamma rays proved to be electromagnetic waves. Cathode rays were shown to be electron beams; channel rays became beams of positive ions; and cosmic rays were later proven to be high-energy particles along with high-energy photons.

Over time, "ray" came to refer only to directional electromagnetic radiation; other phenomena, such as disturbances in nuclear forces, are now termed "waves." (Some papers still refer to "alpha rays" and "beta rays" when they want to indicate the source of radiation, but "alpha particles" and "electrons" are used just as commonly.)

By the late 1800s, it was clear that the world was filled with strange, light-like rays. As these rays increasingly shaped scientists' research agendas, they also captured the imaginations of science-fiction writers.

Herbert George Wells began publishing science fiction in 1895. Within three years, he had already published eight books, including the classics *The Invisible Man*, *The Time Machine*, and *The Island of Doctor Moreau*. For his next work, *The War of the Worlds* (1898), the mighty British Empire was attacked by a technologically superior enemy from Mars. In envisioning a truly alien invasion, Wells produced an amazingly detailed vision that contained grotesque Martians and strange technologies, including Tripod tanks, poisonous black smoke, and weapons that fired invisible "heat rays" that instantly incinerated their targets.[3]

Wells was clearly inspired by the discoveries of the day, particularly that of infrared light. Indeed, in chapter 6, he imagines a kind of infrared searchlight:

> Many think that in some way they [the Martians] are able to generate an intense heat in a chamber of practically absolute nonconductivity. This intense heat they project in a parallel beam against any object they choose by means of a polished parabolic mirror of unknown composition, much as the parabolic mirror of a lighthouse projects a beam of light.

His description was prescient. A century later, those words could be used to explain the operation of an industrial CO_2 laser.

Wells's novel was immensely influential, but it's worth pointing out that there was a predecessor that has been virtually forgotten. In 1809 the American author Washington Irving wrote (as part of his *Knickerbocker's History of New York*) *The Conquest of the Earth by the Moon*, with many of the very same things happening—a superior technological culture from elsewhere in the Solar System invades the Earth with weapons undreamed-of by the inhabitants, and conquer the most technologically advanced nations of the Earth just as the European cultures overtook the nations they invaded and conquered.

> To return then to my supposition—let us suppose that the aerial visitants I have mentioned, possessed of vastly superior knowledge to ourselves; that is to say, possessed of superior knowledge in the art of extermination—riding on hyppogriffs—defended with impenetrable armour—*armed with concentrated sunbeams*, and provided with vast engines, to hurl enormous moon-stones: in short, let us suppose them, if our vanity will permit the supposition, as superior to us in knowledge, and consequently in power, as the Europeans were to the Indians, when they first discovered them. All this is very possible; it is only our self-sufficiency that makes us think otherwise.

In that listing of "Concentrated sunbeams" Irving comes close to Wells's imagining heat rays as an otherworldly and unhuman weapon, without the advantage of all the discoveries of rays of various kinds that occurred between Irving and Wells's time. By the time of Irving's story, however, there were real precursors—Johann Wilhelm Ritter had written of his discovery of "Chemical Rays" (ultraviolet light) in 1801, and William Herschel had written of "Calorific Rays" (infrared light) in 1800. Furthermore, Irving's story was written shortly after François Peyrard had published his work duplicating Archimedes' mirror, which burned things by concentrating sunbeams.[4]

Wells's novel invoked a response in the form of Garrett W. Serviss's 1898 novel *Edison's Conquest of Mars*, in which earth's scientists launch a counterattack on the invaders (described very differently from Wells's Martians) with their own space-traveling ships and armed with a *Disintegrator Ray*—apparently the first time that term had ever been used for a destructive weapon.[5]

Jules Verne didn't like to extrapolate far from known science, but his son Michel didn't feel any such restrictions. After the death of Jules Verne, the son released several previously unpublished works, many of them with changes and additions by Michel. In *The Hunt for the Meteor*, which was published in 1908,[6] the character Zephrim Xirdal creates the "Neutral Helicoidal Ray," which is apparently the first tractor beam to appear in science fiction. This ray can draw objects toward the projector, and Professor Xirdal uses it to bring to earth a meteor made of solid gold. He reverses the beam before the meteor lands, however, after he sees the corrupting influence of people's greed for the gold.[7]

British science-fiction writer George Chetwyn Griffith-Jones, who wrote as George Griffith, specialized in futuristic war stories, depicting aerial battles before the Wright brothers flew, and submarine warfare before it was common. In his 1903 novel *The World Masters*,[8] he describes how Dr. Emil Fargeau, an Alsatian scientist, wanting revenge on Germany, develops a series of inventions that he offers to the French government, but the ministers think he is another crackpot inventor and refuse him. Distraught and penniless, he commits suicide by leaping from a ship. His discoveries, in a waterproof metal box, are found and start a worldwide race to exploit the inventions. An apolitical company is set up that taps the Earth's power using some of the inventions. The devices also include ingenious artillery shells and a method of destroying the cohesion of metal. But the most impressive device is revealed at the end, when a combined French and Russian force attempts to capture the power station:

> For a moment their faces showed white and ghastly in the blazing radiance; and then, to the amazement and horror of those who saw the strangest sight that human eye had ever gazed upon, down the ray of light, invisible, but all-destroying, flowed the terrible energy of the disintegrator on the top of the tower. Their hair crinkled up and disappeared, the flesh melted from their faces and hands. For an instant, two of the most beautiful countenances in Europe were transformed into living skulls, which grinned out in unspeakable hideousness. Then their clothing shriveled up into tinder, and all three dropped together in an indistinguishable heap of crumbling bones. (p. 286)

Elsewhere, besides *disintegrator*, Griffiths calls this weapon a *death-ray*, perhaps the first time that term was ever used. George Griffith returned to disintegrator rays in his 1911 novel *The Lord of Labour*, which describes a future war fought with atomic missiles and disintegrator rays.

The new motion pictures contributed examples, as well—ray guns were photogenic and had immediate impact. In the 1914 serial *The Exploits of Elaine*, the sequel to the immensely popular serial *The Perils of Pauline*, the heroine is menaced by a villain with a *Death Ray*. Coincident with the release of the film was the release of a novelization by Arthur Benjamin Reeve, in what would today be called a "tie-in" (the frontispiece is a still from the serial). In it, some details are given about the Ray:

> LeCroix gleefully patted a peculiar instrument beside him. Apparently it was a combination of powerful electric arcs, the rays of which were shot through a funnel-like arrangement into a converter or, rather, a sort of concentration apparatus from which the dread power could be released through a tube-like affair at one end. It was his infra-red heat wave, F-ray, engine.
>
> The infra-red ray which has been developed by LeCroix from the experiments of the Italian scientist Ulivi causes, when concentrated by an apparatus perfected by LeCroix, an instantaneous combustion of nonreflecting surfaces. It is particularly deadly on the brain centers. It can be diverted, it is said however, by a shield composed of platinum backed by asbestos.

The Viennese villain LeCroix destroys some plants to demonstrate his ray, then kills two people, leaving a coin-sized mark on the forehead. The heroes are able to avoid the ray using their platinum/asbestos shield.[9] Evidently, the inspiration for this "F Ray" weapon draws on the N-rays which René Blondlot of the University of Nancy claimed to have discovered in 1903.

The same team responsible for *The Perils of Pauline* and *The Exploits of Elaine* were back with another alliterative serial the next year, *The Mysteries of Myra*, chapter 15 of which is *The Thought Monster*, featuring a Frankenstein-like creature that threatens the heroine. As if that were not enough, the esoteric cult which set this beast upon her also has Destructive Rays (and protective shields, just as in *The Exploits of Elaine*). The hero manages to wrest away one of the Ray Guns and its shield, and has a duel in the dark with the Master of the Cult, in the course of which the Monster is killed by the ray. Unfortunately, this film is lost.[10]

Another ray-based motion picture from 1916 is the American film *The Intrigue*, in which a young scientist invents an x-ray beam that can electrocute people from a distance. As with Griffith's *The World Masters*, when he makes his findings public and tries to sell them, warring countries vie for the weapon. In 1925 came the Russian science-fiction film *Luch Smerti* ("The Death Ray"), which is reportedly a surprisingly action-filled serial-like adventure, in which the titular ray is used to support a revolutionary rising. The next year saw another Russian work with a death ray, *The Hyperboloid of Engineer Garin* by Alexei Tolstoy (a distant relative of that more famous author Tolstoy). The titular Hyperboloid is a concentrating mirror for light rays, as the Martian Heat Ray of H. G. Wells was. But Tolstoy used the wrong conical section. A Hyperboloid will take rays emanating from a point and make them appear as if they are emanating from another point, or will take already converging rays and make them converge on a different point. To concentrate light you need an Ellipsoid, which will take light emanating from one point and make it converge at another. Or perhaps, as Wells used, a Paraboloid, which will take light emanating from a point and convert it into a collimated beam. The novel was filmed twice in the Soviet Union, in 1965 and 1973.

The weapon in the novel *The Zeppelin Destroyer* (1916) by William le Queux was a ray gun with a surprisingly modest capability—it could cause sparks to manifest themselves on metal frameworks from half a mile away. This, the inventors explain, will be sufficient to ignite any leaking hydrogen gas from the balonets (individual gas bags) inside the zeppelin, causing it all to ignite.[11] Percy F. Westerman's 1923 novel *The War of the Wireless Waves* (written for the juvenile market) pitted British "ZZ" rays against German "Ultra-K" rays. The bad guys in E. E. "Doc" Smith's 1928 proto-space opera *The Skylark of Space* used infra-sound rays, heat rays, ultraviolet rays, and "induction rays" (whatever they might be). Pierrepont Burt Noyes, the son of John Humphrey Noyes, founder of the Oneida community, was a politician and diplomat. In 1927 he wrote *The Pallid Giant*, which tells of how an ancient civilization was destroyed when it discovered a Death Ray that could kill by destroying molecular cohesion, and it ended up completely wiping out that civilization. The story is uncovered just as the French have rediscovered the Death Ray. Here the ray itself is of secondary importance and acts as the catalyst in a cautionary tale about what we would today call Weapons of Mass Destruction. This is probably the most serious use of the concept until this point.[12]

The development in popular culture of ray guns isn't really so surprising—it takes (as popular culture often does) a real observation from the sciences—the plethora of new radiations—Cathode Rays, X-rays, Alpha Rays, Beta Rays, the possible N-rays—and the tense political situation, followed by the First World War, in which new technological developments, including the airplane, the tank, and poison gas, took a part, and extrapolated to a new weapon that fit in with these. Add to this the developing medium of the motion picture, which demanded visually interesting and easily understood images, and the idea of a ray gun was a natural development. Two of the films cited here have battles in the dark—which is where light beams really stand out.

The relentless efficiency of ray guns seems to touch a nerve in readers—then as well as now. Indeed, according to Peter Nicholls's 1979 *Science Fiction Encyclopedia*, the disintegrator "may have resulted from a certain squeamishness, since it allows for a maximum of destruction with a minimum of bleeding pieces to sweep up afterwards."

Ray guns inevitably became staples of what came to be called "planetary romances" and "space operas," but it's notable that they didn't start out that way. Arguably the first planetary romance was Edgar Rice Burroughs's *Princess of Mars*,[13] if we ignore the claims of Edwin L. Arnold's *Lieutenant Gullivar Jones: His Vacation*, and of Gustavus W. Pope's *Journey to Mars the Wonderful World: Its Beauty and Splendor; Its Mighty Races and Kingdoms; Its Final Doom*, both of which have been argued to be inspiration for Burroughs. Burroughs's hero, John

Carter, is a swashbuckling swordsman and general fighter who finds himself mysteriously teleported to Mars. He finds a mix of superscientific advances along with ancient weapons. The Martians have flying craft and atmosphere plants, owing to the operation of the "eighth ray" and the "ninth ray" of the spectrum, but not beam weapons. Burroughs wrote several Mars sequels, but it wasn't until the 1930 entry, *A Fighting Man of Mars*, that he introduced a beam weapon, the *disintegrator gun*. By that time his literary descendants had already armed their romance heroes with ray weapons.

Just as such "Death Rays" were becoming established as a standard item in popular culture—what has come to be termed a "trope" on Internet sites—it suddenly seemed to enter the real world, as several inventors claimed to actually have such "death rays." There are repeated references to them throughout the 1920s and 1930s in newspaper and magazine accounts, which almost invariably called them "death rays." As a group, these inventors tended to be lone, secretive individuals reluctant to say much about their inventions, which (as in the fictional stories) they tried to interest government agencies in buying.[14]

Harry Grindell Matthews claimed in 1923 that he had invented a ray that could stop magnetos in motors, bring down airplanes, and explode gunpowder. He demonstrated some of these capabilities to reporters.[15] Other "Death Rays" were claimed by Edwin R. Scott of San Francisco, Antonio Longoria of Cleveland, Ohio, T. F. Wall of Sheffield University, the famous Nikola Tesla, and a great many others—at least 25 by one count.[16] Reinhard Wulle of the Reichstag claimed that Germany had such a "death ray."[17] It was such claims, presumably, which led the British government to call on Robert Watson-Watt of Bawdsey Research Station to see if Britain could develop such a "death ray" of its own. He presented estimates to show that such a device would be impractical, but offered them a radio detection system, which was the beginning of RADAR.

George Lee Dowd, Jr., "Getting Ready for the Next War," in the December 1927 issue of *Popular Science* observes:

> Not so many scientists and soldiers laugh at the "death ray" as one might suppose. In 1924 a young Englishman named Grindell Matthews came to light as inventor of that. He had lost an eye from emanations of the projecting apparatus. Possibly Matthews has not discovered a workable Death Ray, but that doesn't mean nobody has.[18]

These rays were presented as electromagnetic radiation, or particle beams, or a combination of the two. Despite claims and conspiracy theories to the contrary, none appear to have been satisfactorily demonstrated or put into service. One recent article surveys the field of such rays and their effect on popular fiction.[19]

One more plausible "Death Ray" was the idea of using infrasound to kill, suggested by physicist R. W. Wood in the May 1927 issue of *Popular Mechanics*.[20]

One interesting aspect of the classic ray gun is its sound. Light rays, of course, don't make noise. Infrared and ultraviolet light are as silent as visible light. Particle beams, alpha rays, beta rays, gamma rays, and X-rays make no noise. And in its earliest years, neither did various death rays and ray weapons. Washington Irving's rays, H. G. Wells's, and the other early fictional rays were not described as making any noise. Certainly the silent films required no sound from these beams. But things changed with the coming of radio and sound films. For the radio, especially, some sort of noise was needed, since radio was an aural theater. But even for motion pictures, once sound was available, some sort of signature noise was needed.

And the truth is that, while light itself makes no noise, the means of generating it, manipulating it, or detecting it might not be silent. The x-ray machine in a doctor's or dentist's office has a characteristic whirring sound. The manipulation of large voltages and currents often lead to the crackle of air breakdown or the humming of transformers, and one might expect the associated rays to share this noise. The 1932 film *Sherlock Holmes* featured a very un-Doyle like hero who had superscientific gear, including a Matthews-like weapon for stopping motors. The effects for this were produced by Kenneth Strickfaden, who the previous year had given Universal the electrical equipment used by Doctor Frankenstein to animate his monster, and so this machine had electrical crackling. The word "zap" better describes the noise made by an electrical discharge, like a Jacob's Ladder or a Tesla Coil. Both the sound and the light in such cases are caused by the recombination of ionized nitrogen in the air, rather than by any sort of ray. The word was adopted in the world of science fiction because it adds a satisfying and vital sound to a ray gun's action.

The first use of "zap" in science fiction appears to be by P. F. Nowlan: "Ahead of me was one of the golden dragon Mongols, with a deadly disintegrator ray... Br-r-rr-r-z-zzz-zap." The quote is from Nowlan's story "Armageddon 2419 A.D.," which appeared in the pulp sci-fi magazine *Amazing Stories* in 1928—the same issue, incidentally, that contained the first installment of E. E. Smith's *Skylark of Space* referred to above.[21]

Although Nowlan's works weren't collected and published in book form until the 1960s, they were almost immediately adapted as color comic strips due to their popularity and dramatic visual potential. Nowlan's story of Anthony Rogers, who falls asleep and awakens half a millennium later to find that alien Mongols have taken over North America, translated very well into comics. Rogers joins the underground resistance to fight the Mongols with superscientific weaponry.

The hero's name was changed for the comic from Anthony to the more folksy and plebian "Buck." Buck Rogers was immensely popular, and he inspired similarly named heroes like "Brick" Bradford and "Flash" Gordon; the latter borrowed shamelessly from Rogers, from the Mongols to the faux-Oriental Emperor Ming... right down to the zap!

"Zap," by the way, appears to have started entering the nongenre popular consciousness by the early 1940s, where it was roughly synonymous with "hit." It shows up in a 1942 song title "Zap the Jap, Rip up the Nazis," and appears in *Billboard* magazine the same year.

The heat ray used in the 1938 Mercury Theater on the Air Halloween Broadcast of *War of the Worlds* used a somewhat electrical sound, but a smoother, more continuous one than the staccato "Zap." Other uses of the ubiquitous ray invented completely new sounds. The very first Superman cartoon, released on September 26, 1941, had a Mad Scientist who threatens Metropolis with his "Electrothanasia ray" unless a ransom is paid. It turns out to be a white laser-like beam projected from what appears to be a gun turret mounted in an observatory. Its sound is a whine, and when Superman interposes himself between the ray and the base of the building it is destroying, he physically reels backwards. He flies toward the source of the beam, punching away at the ray as if it were a physical adversary—which it acts like, pushing him backwards. It all seems silly—why not simply go around the beam and strike directly at the observatory housing it? But that would not be sufficiently dramatic—and drama is what's called for, not verisimilitude. Superman eventually gets to the barrel of the ray cannon and ties it in a knot. The titanic energies, with nowhere to go, begin to back up, overheating first the barrel and then the guts of the machine, which eventually bursts.

In George Pal's 1953 film of *War of the Worlds*, the Martian heat-ray spews out a pulsed series of falling tones that has become iconic and has been reused in other films (such as 2004's retro-SF *Sky Captain and the World of Tomorrow*), pulsed bursts characterized the "blasters" in 1956's *Forbidden Planet* as well. But it was becoming more usual for the continuous ray to sound like a continuous tone, as in *The Day the Earth Stood Still* (1951), *This Island Earth* (1955), *Earth vs. the Flying Saucers* (1956) (which also gave Earth people a ray weapon to counter the aliens'), and the Japanese *The Mysterians* (1957). Single tone, possibly modulated, became the standard "beam weapon" sound. This was exemplified in the "phasers" of *Star Trek*.[22]

Another sound was a chirp, which could be easily generated by the sound crew placing a pickup on a stretched cable (like the cable restraining a telephone pole) and striking it on the side. This gives a futuristic melodic burst that belies its pedestrian origin. *Star Trek* used it for its "photon torpedos," mortars, and the like. Later, *Star Wars* used a similarly generated sound.

The use of sound fills a visceral need—we like to think that *something* is happening to verify that titanic energies are in play. Robert Sheckley made a point of this in his 1958 short story "A Gun without a Bang"—animals didn't associate the destructive power of a ray gun with the gun itself when it made no noise. It's not clear to me that animals associate the destructive power of a noisy gun with its noise, either. But certainly *people* do, and a proper powerful weapon ought to indicate its potency with sound. So when Goldfinger threatens James Bond with the latest in weaponry, an industrial laser, in the movie *Goldfinger* (it was a circular saw in the book), the fiendish device starts with an ominous whipcrack, followed by a high-pitched trilling whine as it cuts. When the TV show *The Avengers* (in the episode *From Venus with Love*) showed a powerful, possibly extraterrestrial ray, it was accompanied by a revving-up whine (which proves its undoing, as a scientist later identifies the sound as that of a "laser"—although this is evidently only characteristic in that particular television reality).

Another feature of ray guns is directly related to the medium. Whether shown in drawings or on the screen, the audience has to realize that the beams are present, where they are coming from and going to, and the direction they are proceeding. This has led to the completely unphysical (but highly dramatic) depiction of rays visible from the side, as if they were being scattered by a plethora of dust motes in the air (and, in truth, if it were a very intense beam, it might become visible in this way),[23] and of its progress being visible as it moves from source to target. Often the animation used to produce these moving beams is so rapid as to be almost "subliminal" (if you'll pardon the pun), but the "blaster" beams in *Forbidden Planet* are in staccato bursts that the eye can clearly and unphysically follow. The phaser beams from *Star Trek* are clearly visible moving along. And both the *Forbidden Planet* and *Star Trek* beams "splash" prettily but unrealistically when they rebound from some uncooperative target.

Notes

1. P. F. Dahl, *Flash of the Cathode Rays: A History of J. J. Thomson's Electron* (1997), 49–57.

2. E. H. Kennard et al., *Introduction to Modern Physics*, 5th ed. (1955), 455.

3. One of the most interesting details is the way that Wells also depicted the Martian *lack* of technology—they evidently did not have the wheel, like the indigenous American cultures before Columbus. Thus, the Martians have wheel-less tripods instead of wheeled vehicles. In the 2005 Spielberg adaptation, they hint at this when the (definitely non-Wells) Martians go through a cellar and are interested in the operation of a bicycle wheel.

4. See chapter 2, "The Solar Weapon of Archimedes."

5. The term "disintegrator" had been used seven years earlier in the novel *The Disintegrator: A Romance of Modern Science* by Arthur Morgan and Charles R. Brown. The disintegrator of the title does take things apart down to its constituent atoms, but as one half of what we would today call a teleportation apparatus, the other half being, in the novel, the "re-integrator." This is by no means the first teleportation story, which dates back at least to Edward Page Mitchell's 1877 *The Man Without a Body*. Prior to this *disintegrator* meant essentially a grinder, which could pulverize solid material to fine powder and was a perfectly normal and acceptable mining term.

6. *The Meteor Hunt*, trans. and edited by Frederick Paul Walter and Walter James Miller (2006).

7. The original novel, without Michel's additions, was finally released in 1988 and an English translation published in 2006. The oddly-named Zephrim Xirdal and his wonderful rays are completely missing—and the story works quite well without him.

8. George Griffith (George Chetwyn Griffith-Jones), *The World Masters* (1903). Available online at http://books.google.com/books?id=sjogAAAAMAAJ&pg=PA298&dq=%22death+ray%22&hl=en&sa=X&ei=fSGDT6mpFrSyoAGA3_D7Bw&ved=0CHQQ6AEwCQ#v=onepage&q=%22death%20ray%22&f=false, accessed April 9, 2012.

9. The novelization is available at several Internet sites, including this one: http://books.google.com/books?id=owBUAAAAYAAJ&printsec=frontcover&dq=Exploits+of+Elaine&h

l=en&sa=X&ei=DCOCT8vzBKbgoQGvxdmeCA&ved=0CDYQ6AEwAA#v=onepage&q=Exploits%20of%20Elaine&f=false, accessed April 8, 2012. The quotations come from pages 173–174 and 182. I am grateful to Jeff Hecht for information about this appearance of a beam weapon, which he incorporated into his article "Half a Century of Laser Weapons," *Optics and Photonics News* 20 (2): 14–21 (February 2009).

10. This description comes from page 1299 of the journal *Moving Picture World*, August 19, 1916. There is also a recent "reconstruction" using script and stills, *The Mysteries of Myra* by Charles Goddard and Eustace Hale Ball, *The Serial Squadron* http://www.serialsquadron.com/, accessed April 9, 2012.

11. Available online at http://openlibrary.org/works/OL810491W/The_zeppelin_destroyer, accessed April 9, 2012.

12. E. F. Bleiler, *Science Fiction: The Early Years*, 567: http://books.google.com/books?id=KEZxhkG5eikC&pg=PA567&dq=The+Pallid+Giant&hl=en&sa=X&ei=24CLT_KuO-nhoQGy1tXOCQ&ved=0CEcQ6AEwBA#v=onepage&q=The%20Pallid%20Giant&f=false, accessed April 15, 2012. Quotes from *The Pallid Giant*: http://books.google.com/books?id=Uh83AAAAIAAJ&q=%22Death+ray%22&dq=%22Death+ray%22&hl=en&sa=X&ei=tX-LT5WZC8SEoQG30MTDCQ&ved=0CFMQ6AEwBg, accessed April 15, 2012.

13. Originally serialized in 1911 as *Under the Moons of Mars*, as by "Norman Bean." Burroughs really wanted the pseudonym to be "Normal Bean," meaning "Normal Being," but the typesetter or editor apparently didn't see the joke, or perhaps didn't care for it.

14. "Death Rays" thus seem to belong to that class of invention that was influenced by popular culture depictions and descriptions. The "Atomic Bomb" is the classic case of this. Leo Szilard admitted that his interest in the idea and its political significance was initiated by the description of "atomic bombs" in H. G. Wells's 1914 novel *The World Set Free*, which depicts such weapons used in war, dropped from airplanes on the cities of the opposing side. Wells got his inspiration, in turn, from the works of physicist Frederick Soddy. A more relevant case is the TASER, the electric stun-gun used by police forces. I always assumed that the name was given to this device in imitation of the *LASER*, but inventor Jack Cover, a former NASA researcher, has said that he conceived of the device and named it after a similar one in the 1911 boys' novel *Tom Swift and his Electric Rifle*—the name is an acronym for Thomas A. Swift's Electric Rifle. He had to "cheat" a bit for the middle initial, since Tom Swift's middle name was never given.

15. See "Death Ray is Carried by Shafts of Light," *Popular Mechanics* 42 (2): 189–192 (August 1924), which contains some impressive drawings of the claimed capabilities of the ray, along with a picture of Matthews and the motorcycle engine he claimed to have stopped. "Invisible Death," *Time*, April 21, 1924.

16. http://airminded.org/2009/03/27/the-death-ray-men/, accessed April 16, 2012. Here's a (partial) list of others who claimed to have invented a death ray before 1939:

(a.) unnamed chemist, Bradford, 1916

(b.) Wulle, a "militarist" Reichstag deputy, claimed that Germany had a death ray. Presumably Reinhold Wulle

(c.) Grammachikoff, Soviet Union

(d.) unnamed engineer, Paris

(e.) "a German at the radio station at Nauen"

(f.) unnamed inventor, France

(g.) Bernays Johnson, United States

(h.) Philipoff, editor and publisher of *Scientific Review*, Soviet Union

(i.) unnamed, Tunbridge Wells

(j.) unnamed man, Manchester

(k.) Dr. T. F. Wall, electrical engineer, Sheffield University, 1924

(l.) Edwin R. Scott, San Francisco, 1925

(m.) Henry Fleur, San Francisco, 1936

(n.) "Professor Anthony—an M.A., M.D., D.Sc., Hon. Professor of Natural Science and Philosophy, and holder of degrees in English Botanic Medicine," 1937

(o.) R. Russell Clarke, barrister and Room 40 cryptographer, 1917

(p.) Coxhead, Maidenhead, 1933

(q.) Ulivi, Italy, 1913

(r.) unnamed corporal, (British) 4th Army, c. 1916–9 (*The Times*, 14 October 1937, 14)

(s.) Prior, Britain, 1924

(t.) Raffe, Britain, 1924

(u.) unnamed German inventor, represented by British engineer John H. Hamill, 1924

(v.) Nikola Tesla, United States, 1934

(w.) Dr. Alberto Longoria, United States, 1934

(x.) Henri Claudel, France, 1935

(y.) Prof. Harry May, Britain, 1936

(a.) through (n.) are from E. H. G. Barwell, *The Death Ray Man: The Biography of Grindell Matthews, Inventor and Pioneer* (n.d. [1943]), chapter 16; (o.) and (p.) are from David Zimmerman, Britain's Shield: Radar and the Defeat of the Luftwaffe (2001), 45–47. The others are sourced as indicated. The dates given are usually when the claim was made public, though in some cases it's when the invention took place. Some of these are no doubt duplicates—Barwell doesn't give many details in most cases—on the other hand, there are no doubt many more names still to be found. In any case, it's clear that a death ray was much sought after in both Europe and America in the 1920s and 1930s. And if there was anything to it, you'd think that one of these inventors would have produced a working example, instead of just a mass of press clippings.

I note as well that *Popular Science* for January 1935, p. 17, reports on a French inventor of a "light ray gun," which can blind and incapacitate animals and people. It appears, from the description and picture, to be a flash gun with a concentrating parabolic reflector attached. http://books.google.com/books?id=lyoDAAAAMBAJ&pg=PA17&dq=%22Ray+Gun%22&hl=en&sa=X&ei=PMaNULnnK-XZoQGMw4GoCA&ved=0CFsQ6AEwCQ#v=onepage&q=%22Ray%20Gun%22&f=false, accessed October 27, 2012.

17. Previous note and "The Death Ray Rivals," *New York Times*, May 29, 1924.

18. George Lee Dowd, Jr., "Getting Ready for the Next War," *Popular Science* 111 (6): 26–27, 168–170, especially p. 27 (December 1927).

19. William J. Fanning, Jr., "The Historical Death Ray and Science Fiction in the 1920s and 1930s," *Science Fiction Studies* 37 (2): 253–274 (July 2010).

20. R. W. Wood, "Can Inaudible Sounds Kill?" *Popular Mechanics* 47 (5): 705–706 (May 1927).

21. The cover, showing a levitating suited man, illustrates the Smith story, but is often misidentified as illustrating the Nowlan story. Note: *The Oxford English Dictionary* lists the first appearance of "zap" as May 7, 1929, in the *Washington Post*, but citing the same line by Nowlan—clearly a quote from his earlier work. (2d ed., Vol. 20, p. 792 (1989), entry on Zap). I note that the first appearance of "ray" in the sense we use it for "the transmission of energy" is given by the same edition as 1664 and the first "science fictional" use as H. G. Wells, *War of the Worlds* in 1898 (see Vol. 13, p. 240, entry on *Ray*).

22. Creator Gene Roddenberry said that he originally called the beam weapons of *Star Trek* "lasers," but changed it to the similar-sounding "phasers" because he was afraid of technology catching up with the show and not being capable of doing what his nominal "lasers" did. A good move, too. None of the lasers I ever worked with were capable of the "neural impact" that was the "stun" setting.

23. About the only one to try to excuse this side-visibility on technical grounds was Arthur C. Clarke, in his novel *Earthlight*.

{ 37 }

Mystic Cameras

At the beginning of the 1974 Mel Brooks film *Young Frankenstein*, Gene Wilder, playing Frederick Frankenstein, the grandson of the creator of the famed Monster, is lecturing a class at medical school. One impertinent Young Spark asks, "Isn't it true that Darwin preserved a piece of vermicelli in a glass case, until by some extraordinary means it began to move?" Frankenstein (*Frohnn-kon-steen*, as he insists it be pronounced) wishes to deflect any comparison with his notorious grandfather, deflects this by sarcastically asking, "Do you mean the *worm*, or the *spaghetti*?" Undeterred, Bright Spark replies, "Why the *Worm*, of course."

Mel Brooks and Wilder, who together wrote the screenplay, certainly did their homework. That question of the student's is almost word for word from Percy Shelley's preface to the 1818 edition of his wife's famous book. Even so, I was surprised to learn that they got it wrong. There is no worm called "vermicelli"—Wilder and Brooks evidently confused the name with *vorticella*, which *is* a worm. Their confusion was helped by Shelley's assertion, which seems to be a confused mixing of two findings by Erasmus Darwin (grandfather of evolutionist Charles). He observed an apparent case of spontaneous generation of eel-like creatures from a paste of flour and water, and on another occasion noted that vorticella can apparently come back to life after being dried if they are soaked in water.[1] This and other confused recollections of science were extrapolated to make the background for the story of the creation of life. One of the other factoids marshaled to this purpose are the experiments of Luigi Galvani, where an electric spark was found to excite movement in the muscles of a frog's leg. This idea is barely hinted at in Mary Shelley's book, but it became a centerpiece of the cinematic adaptations of the work, so that today our standard image of the Mad Scientist includes huge discharge-shedding electrical apparatus. *Frankenstein* was by no means the only example of electricity being used as the agent of bringing life. Edgar Allan Poe used it in three different stories.

The idea that a new and powerful force can be used to produce mysterious effects (often far beyond what it can actually accomplish) is a common one in literature and folklore. Similar things have been alleged of hypnotism, x-rays, cosmic rays, radioactivity, and, most recently, genetic engineering, and nanotechnology. At one time, it may surprise you to learn, it was alleged of photography. The idea that "a man commanded the sun to paint his portrait, and it did" (as Poe put it in *The Thousand and Second Tale of Scheherazade*) at one time seemed utterly amazing and mysterious, and was viewed with much amazement and awe. There were stories of mysterious photographs produced by nonhuman means.

There were many reports from the late nineteenth century of "photographs" being made on window glass by the action of lightning. No lens or imaging apparatus was involved, and the image was not a silhouette.[2] In a typical case, a person who died during a thunderstorm, or who was nearby when lightning struck has their features preserved on the glass as if by photography. In one account, an old woman died after lightning struck a tree near her "8 x 10 window." "When the neighbors went to lay her out... they discovered... that on one of the panes of glass in the window against which her bed stood was a perfect photographic likeness of her as she appeared in her neat cap and gown." Other accounts may have the image preserved on a mirror.

As Folklorist Barbara Allen notes, all of these reports come from local newspapers in the 1870s to 1890s. Reports of such "miraculous" photographs have evaporated, but there continue to be reports of suspiciously similar religious imagery appearing in glass and other media. Contrary to the reports, none of the lightning photographs have been examined by baffled scientific investigators.

It seems pretty obvious that these "magic photos" are examples of Urban Legends from over a century ago. But why did they emerge when they did, and then so rapidly die away? Photography had been around since the 1830s. Allen argues that "the introduction of glass plates in 1851... made photography a truly democratic invention." But, although common, the mechanics of photography remained mysterious and vague to the same Americans who were its customers. The photographer snapped the portrait, then disappeared into his wet chemistry lab to perform odd manipulations on his glass plate. Add the use of magnesium flash powder to illuminate an indoor scene, and you have something not far removed from the reported process.

The introduction of celluloid as a base for the photographic emulsion in the 1880s, claims Allen, broke the association of photography with glass in the popular mind, and stories of miraculous glass photos disappeared from the newspapers.

Another series of stories revolves around miraculous images appearing on the retinas of the eyes. In its full-blown form, the last sight of a murdered person can be recovered from a dead person's eyes and used to track down the criminal.

Such an odd, technology-based belief must be of recent origin and has obvious similarity to the Lightning Photo stories.[3] The earliest known account of this idea occurs in 1857. The magazine *Notes and Queries* cites the *New York Observer* which, in turn, reports from an undisclosed English newspaper "that the last image formed on the retina of the eye of a dying person. Remains impressed upon it as on a daguerreotype plate. Thus it was alleged that if the last object seen by a murdered person was his murderer, the portrait drawn upon the eye would remain a fearful witness in death to detect the guilt, and lead to his conviction."

Where did such an idea come from? The mechanics of vision has been understood since Johannes Kepler in the seventeenth century. Christoph Scheiner was able to verify Kepler's theory of imaging upon the retina by dissecting cow eyes. Evans notes that these reports follow the invention, in 1850, of the ophthalmoscope by Helmholtz, which allowed direct inspection of the retina.

An 1863 French report claims that an English photographer was able to obtain a photograph from the eye of a steer shortly after its death, which showed "details of the cobblestone floor of the slaughterhouse." The idea was so common that Villiers de l'Isle-Adam used it in a short story, "Claire Lenoir," in 1867. The story first recounts as fact that the eyes of animals recently killed retain images of their place of death, then goes on to describe a case in which such an image is retained in a human eye.

All of this, despite the assurances of scientific accuracy, it must be stated, was utterly without any real scientific backing. That was to change in 1876, when Franz Boll at the University of Rome isolated the biological pigment associated with sight, which he called *visual red*, but which soon came to be called *visual purple*, or *rhodopsin*. His colleague in Heidelburg, Wilhelm Friedrich Kühne, soon absorbed himself in the study of rhodopsin and performed experiments that resembled those reported in the press. He actually did allow the eyes of an albino rabbit to dark-adapt (insuring a large stock of rhodopsin), then made an exposure of several minutes of a scene of a barred window by forcing the rabbit to stare at it. He then executed the rabbit, removed and dissected the eyes, and "developed" the retina in a solution of alum. The result was a biological photograph that showed the clear image of windowbars. Kühne called it an *optograph* and proceeded to make more.

It's as remarkable as the development of the laser as in "validating" the science-fiction notion of the Ray Gun, as first suggested in H. G. Wells's *War of the Worlds* (and Washington Irving's lesser-known *The Invasion from the Moon*). A decade after the notion first appeared in print, Kühne demonstrated the reality of preserved retinal images.[4] As an unintended by-product, it also provided a reason for people to believe that there was truth in the notion that one might catch a murderer through postmortem studies of retinas.

In fact, there was little real basis for the belief that the retina of a murder victim would retain useful information. Aside from the vast difference between the careful preparation of Kühne's test animals and their rapid dispatch immediately after viewing, and the case of haphazard human victims examined hours or days later, there were, in fact, actual tests of the notion. Kühne himself excised and treated the eyes of a criminal who had been guillotined, but obtained no discernible image. Furthermore, in 1869 the Society of Forensic Medicine in France had Dr. Maxime Vernois study the problem. His report, issued in December of that year and subsequently published, saw no utility in the practice at all.

Nevertheless, the idea was too powerful an idea not to be used in fiction. Rudyard Kipling uses the idea of images from a dead man's eyes in his 1891 story "At the End of the Passage." Jules Claretie used it in his novel *L'Accusateur: L'Oeil du Mort* in 1897. Cleveland Moffet employed it in his 1900 short story "On the Turn of a Coin." Jules Verne used it as the climax of his novel *Le Frères Kip* in 1902.[5] Thomas Dixon's 1905 novel *The Clansman* (upon which the film *Birth of a Nation* was based) uses it as well. James Joyce mentions it in passing in *Ulysses.*

Campion-Vincent's article lists several more recent works in which the concept is used or mentioned, but it came out too early to include one of the grislier appearances in recent years. In the 1999 film *The Wild, Wild West* light is projected through the backs of a dead man's retinas as if they were biological slide projectors, throwing his last sight upon a wall. More recently, Derek Ogbourne has put together a traveling exhibition and anthology, *The Encyclopedia of Optography: The Shutter of Death.*[6]

Notes

1. Mary Wollstonecraft Shelley, *The Annotated Frankenstein*, annotations by Leonard Wolf (1977), 4.

2. I first learned of this from Frank Edwards's 1959 book *Stranger than Science*, named after his syndicated radio show. Chapter 62, "Fantastic Photos," contains his uncritical recitation of these accounts. The same reports were examined more critically (and accurately—Edwards gets the dates wrong) by folklorist Barbara Allen in "The 'Image on Glass': Technology, Tradition, and the Emergence of Folklore," *Western Folklore* ***41*** (2): 85–103 (April 1982).

3. See Allen's work, cited above. A fuller account appears in Veronique Campion-Vincent, "The Tell-Tale Eye," *Folklore* 110: 13–24 (1999). Another is Arthur B. Evans, "Optograms and Fiction: Photo in a Dead Man's Eye," *Science Fiction Studies* **20**: 3, No. 61: 341–361 (November 1993), available online a several places, including http://jv.gilead.org.il/evans/optogram.html, accessed September 30, 2012.

4. Kuhne's work was reproduced 30 years ago by E. Alexandris and T. Klothermann, reported in "Optography of the Retina: Resumption of Kuhnes Studies," *Ber Zusammenhunst Deutsch Ophthalmol. Ges.* ***74***: 567–568 (1977).

5. This novel was published in English translation for the first time in 2007. *The Kip Brothers*, trans. Stanford L. Luce, ed. Arthur B. Evans (2007).

6. http://www.derekogbourne.net/www.derekogbourne.net/Home.html, accessed September 30, 2012.

{38}

Playing with Light

The December 1941 issue of *Astounding Science Fiction* magazine contained the short story "Bullard Reflects." It was written by Malcolm Jameson, a retired Navy man who took up writing science fiction when cancer forced him to do less strenuous work. The story is the fourth in a series about Captain Bullard of the space ship *Pollux*, part of the Space Patrol that kept order throughout the Solar System. It begins with a description of the game of *Dazzle Dart* between the *Polliwogs* of the *Pollux* and the *Beans* of the sister ship *Castor* Each team's "quarterback" has a "superflashlight" equipped to produce one hundred flashes of light, each ten seconds long. The other team members have slightly concave mirrors attached by straps to arms and legs, and the object is to bounce the one centimeter diameter light beam from flashlight to mirror to mirror and eventually into the team's goal. Opposing team members can block the beam, or even try to redirect it into their own goal. That the game is played in the lower gravity of the moon adds to the strategic possibilities.[1]

Jameson's game is interesting and well thought out. The ten second burst gives the players time to redirect the beam without having to anticipate the conditions. The slightly concave mirrors act to counteract the beam spread. The only major change one would make would be to substitute a laser perhaps, an option not available to a fiction writer in 1941. But for some reason the game never caught on, even among science-fiction fans noted for recreating and playing fictional games such as Edgar Rice Burroughs's Martian Chess, *Jetan*.

Why not? Certainly the technology existed then to play such games with light. The technology to play relay games with light beams arguably existed as far back as ancient Egypt, where some people have theorized that polished metal mirrors might have been used to relay sunlight into tombs. Whether they did or not, I'll bet someone used a mirror to beguile his cat the same way I do mine with my laser pointer.[2] But I am aware of no such cooperative passing games using light. In fact, there have been few games in which light plays an essential part.

The end of the nineteenth century would have been the ideal time for such games to begin. The electric light had been invented and before the end of the century was made portable by batteries and generators, and the selenium photocell had been invented and had been used by Alexander Graham Bell to transmit telephone signals on light beams. But no Light Games appeared until 1935.

The *Rayolite* company was founded in Tulsa, Oklahoma, in 1934. In April of that year, Charles W. Griffith and the *Rayolite Rifle Range Company* filed a patent for a game using mechanical and optical elements. A device shaped like a rifle fired a beam of light at a target that bore a corner cube reflector. The "rifle" barrel also hosted a photocell detector that registered a voltage when the light reflected from the corner cube back in the barrel. The patent was granted on July 2 of the following year, and advertisements for the device appeared immediately. The games were manufactured by the J. P. Seeburg Corporation of Chicago and sold for $495 apiece. It cost five cents for a set of 10 to 20 shots at a moving target. Games were placed in the same places that pinball machines, and, later, video games would be placed—bars, small restaurants, and arcades.[3]

The game was apparently very popular, and a series of new patents supported many variations for the game. Eventually, players had the options to shoot at ducks, bears, raccoons, jailbreakers, and (during the Second World War) war targets. Other companies soon deduced how to build light games without infringing on the *Rayolite* patents, and these games continued to be produced until the video revolution substituted video games for mechanical ones.

With Cathode Ray Tubes as displays, the light guns no longer had to provide their own light source, but could simply detect the light from the CRT itself. Originally they could detect which target (if any) was being pointed at by temporally correlating the applied signal with the detected one. For current plasma and LCD displays more refined detection schemes are used.

The next step in the evolution of light games was the United States Army contracting for the development of MILES—the Multiple Integrated Laser Engagement System—in the 1970s. This system for simulated warfare used a laser module that was attached to an actual gun barrel. When the gun fired a blank cartridge, the laser fired a coded burst that could be received by a sensor worn by the opponent and would register a hit only if the shooter was, indeed, on target. The system has continued in use and was upgraded in the 1990s. A related system, the Engagement Simulation System (ESS) was used by the Department of Energy to train the police forces used to guard nuclear power plants. The MILES system helped to inspire civilian counterparts.[4]

In 1979, coincident with the release of *Star Trek: The Motion Picture*, South Bend Toys of Indiana produced *Star Trek Electronic Phaser Guns*. These used infrared sources and detectors, and registered hits from the other gun with "phaser" noises. Five years later, George Carter III of Texas produced *Photon*, a game he conceived of while watching the movie *Star Wars* in 1977. His game involved light guns and individually worn receivers, and he set up an arena for teams equipped with such guns and receivers to fight it out with each other. The first *Photon* center opened in Dallas in March of 1984. A franchise opened in Toronto the following year. A year later *Worlds of Wonder* produced *Lazer Tag*, which used a similar format, and within a short time there were many imitators. The market for these slowed in the early 1990s, but has more recently regained some force.[5]

These games, from *Rayolite* onward, all are simple shooter games, with photons substituting for solid bullets. There was nothing more exotic than a simple direct line from source to target, with no ricochet or reflection involved. Intriguingly, something closer to Jameson's vision did appear in the virtual world. In the 1970s a simple computer game version of *Dazzle Dart* appeared, using the very same name for the game as in "Bullard Reflects." The game was scaled to the modest capabilities of the computers of the period, with very few players per side, limited mobility of players and mirrors, and a two-dimensional Board. Nevertheless, the game allowed for the first time for shots and reflections.[6]

No one seems to have yet directly updated *Dazzle Dart*, but other cyber games with reflection have appeared, such as *Yet Another Laser Game*.[7]

In the real world, people have suggested using laser pointers and mirrors to direct a beam onto a target. These suggestions seem to have been made by those ignorant of Jameson's story.[8] But these suggestions have not become reality, or at least widespread reality. Possibly this is because the unrestricted use of lasers—even laser pointers—is discouraged for safety reasons. No one wants a stray laser beam in the eye. And the use of protective goggles would prevent the shooter from seeing where his beam was going. There are ways around this—using, for instance, a short wavelength source and fluorescent targets—but nothing along those lines has been done. I suspect that, were one to follow Jameson's description, a winning strategy would be to rapidly sweep the beam across the target area in a sort of raster scan, rather than carefully aligning for a shot. The Game Designer could control for this by requiring a "dwell" time on the target before the signal registered.

Arguably the closest anyone has come to the idea of a reflecting laser game in the real world is *Khet*, a chess-like game developed by Prof. Michael Larson of Tulane University, along with his students Del Segura and Luke Hooper.[9] It was originally done as a class project, but was soon introduced as a commercial toy in 2005 under the name *Deflexion*. The name was changed to *Khet* on September 15, 2006, when the game was given an Egyptian flavor. The pieces, which can be placed and moved on squares on the board, may have mirrors, or ordinary surfaces, or, in the latest variation, beamsplitters. The object is to manipulate the laser with one's mirrors so as to strike the opponent's "king," called a "Pharaoh." The laser gimmick is essential to the game, and the use of well-defined pieces and lasers in a fixed plane eliminates the issue of eye safety, so long as players are careful and responsible.

This is the closest anyone has come to Jameson's game, or any sort of light-based game. It's surely not the closest or the cleverest, however. There's no reason one couldn't build a laserless real-life version of *Dazzle Dart* and play it. And there are surely other possible Light Games waiting to be discovered. Imagine teaching undergraduates about polarization states using a laser game fitted with linear and circular polarizers, half- and quarter-wave plates, Fresnel rhombs, and dielectric interface reflections.

Since this originally appeared, there have been other incarnations of laser games, and I've been made aware of others. The Trumpf laser company of Ditzingen, Germany, built a tabletop laser game that resembles Jameson's *Dazzle Dart* game in many respects, and displayed it at various locations in 2011, including the Boston Museum of Science.

One student built a tabletop laser chess game using resin pieces as part of a course and posted it on his blog in 2012.[10]

A computer version of such a game (with no actual lasers involved) appears at the site *Arcade Street*.[11]

In 2009 Professor Ellen Yi-Luen Do and two graduate students from Georgia Institute of Technology, David Joyner and Chih-Sung (Andy) Wu, presented a paper at the Eighth International Conference on Interaction Design and Children (IDC 2009) in Como, Italy, on *Tangible Optical Chess: A Laser Strategy Game on an Interactive Tabletop*.[12] In this case the game started as a GUI on a computer, but they advanced to a tangible but not "practical" model that could be manipulated by hand, although no real lasers (or mirrors) were used. The game was on a "Tangible TableTop" (TTT), with the beam paths projected onto the table. This got around the possibility of the laser beam being inadvertently directed into someone's eye.[13]

Notes

1. Malcolm Jameson, "Bullard Reflects," *Astounding Science Fiction* (December 1941), reprinted several times, including *A Treasury of Great Science Fiction*, ed. Anthony Boucher (1959).

2. See chapter 34, "This is your Cat on Lasers."

3. On the Seeburg RayoLite and similar games, see the *Sands Mechanical Museum* site online at http://www.sandsmuseum.com, accessed July 23, 2007; the Pink Godzilla site at http://www.pinkgodzillagames.com/retro_reviews/the_early_history_of_the_light_gun.php, accessed July 23, 2007 (and I'm not responsible for what they call themselves); and http://marvin3m.com/arcade/rayolit.htm, accessed July 23, 2007.

4. On MILES, see http://www.fas.org/man/dod-101/sys/land/miles.htm and http://home.comcast.net/~Ferret1963/All_Systems.HTML, both accessed July 23, 2007.

5. On *Star Trek* Phasers, Photon, and Lazer Tag, see http://home.comcast.net/~Ferret1963/All_Systems.HTML, www.lasertag.org/general/history.html; and http://en.wikipedia.org/wiki/Laser_tag, accessed July 23, 2007.

6. The only reference I've been able to locate on the computer version of *Dazzle Dart* is at http://www.kaleberg.com/dazzle/dazzle.html, accessed July 23, 2007. *Yet Another Laser Game* is at lhttp://www.gamextazy.com/games/logic-puzzle'yet-another-laser-game.htm, accessed July 23, 2007.

7. *Yet Another Laser Game* is published by Persistent Realities "YALG—Yet Another Laser Game is © copyright 2002–2004 by Almar Joling" See http://persistentrealities.com/yalgsite/ or http://yet-another-laser-game.down724.com/, accessed October 21, 2012.

8. For such suggestions, see "Laser Goal" at The Halfbakery http://www.halfbakery.com/idea/Laser_20Goal, accessed July 23, 2007; and "Laser Pointer Games" at Cool Stuff Archives http://www.arborsci.com/CoolStuff/CoolSuffIII.htm, accessed July 23, 2007.

9. There are many sites on *Khet* on the Internet. A useful one, with history, is at http://en.wikipedia.org/wiki/Deflexion, accessed July 23, 2007. The official site is http://www.khet.com/.

10. Alex Van de Sande's *Wandering About* blog at http://wanderingabout.com/portfolio/laser-chess/posted 2012, accessed April 23, 2012. Sadly, he said that his professor wasn't impressed, although his friends were.

11. http://www.arcadestreet.com/laser-prisms.htm, accessed April 23, 2012.

12. http://synlab.gatech.edu/data/papers/joyner_idc2009_optical-chess.pdf, accessed April 23, 2012, later published as a paper, "Move, Beam, and Check! Imagineering Tangible Optical Chess on an Interactive Tabletop Display," *ACM Computers in Entertainment* 8 (3): Article 20 (December 2010). Available online at http://theauk.net/media/pdf/CiE_Wu_2010.pdf. See also http://synlab.gatech.edu/projects/opticalChess/ and http://acmelab.gatech.edu/?p=3321, accessed April 23, 2012.

13. The Trumpf model used rotatable "men" that were within a sealed Plexiglas box, so the beam was kept in a single plane inaccessible to the eye, but the other games listed here do have a potential for directing the beam out of the plane of play.

{39}

I Must Find That Tractor Beam

When he died in 1905, Jules Verne left several unpublished stories that his son, Michel, edited and subsequently put into print. One of these was *The Hunt for the Meteor* (*La Chasse au Météore*), which appeared in 1908, three years after his death. Verne's original story, inspired by his own novel *Hector Servadac* (translated as *Off on a Comet*) related the consequences of the fall of an unusual meteor and its unexpected effects on human society. The meteor, it turns out, is made of gold. Verne saw interesting possibilities in the situation, which was, according to the science of the time, plausible, if far-fetched.

As was the case with many of Verne's posthumously published works, however, this one shows the hand of his son. Michel Verne was much more flamboyant in his speculations and wasn't limited to straight extrapolation and plausibility.[1] He added chapters to his father's work, introducing an eccentric hermit scientist with the unlikely name Zephirin Xirdal, who invents the *neutral helicoidal ray* to pull on the meteor and alter its course to bring it down to earth. Later he invents a ray of *neutral rectilinear currents* to push on the falling meteor and can thus adjust its course.

Such scientific gobbledegook was not to the elder Verne's taste—he liked to have his stories grounded in demonstrated science. He took H. G. Wells to task over his use of the nonexistent anti-gravity metal *Cavorite* in *First Men in the* Moon.[2]

But the idea had been sown. Rays for propulsion, pushing away things, and allowing ships to rise were used in Edgar Rice Burroughs's John Carter of Mars stories, and in Philip Nowlan's *Anthony (Buck) Rogers* stories, but beams that pulled on an object took a little longer to catch on. Edmund Hamilton introduced such an attractive beam in his novel *Crashing Suns*, serialized in the magazine *Weird Tales* in August and September 1928, and published later that year in book form. The novel concerned what the title literally implied—suns that were on a collision course. As with Michel Verne, Hamilton introduced the concept of a nonphysical tether to do what a physical tether could not do—grab a rapidly moving and very distant (and in this case, extremely hot) celestial body. You could argue that the necessity of the story drove the need to invent such a device.

This was not the case with most subsequent users of the device. Edward E. "Doc" Smith started using them in his 1931 story of his "Skylark" series, *The Spacehounds of IPC*, in which the ray is used to grapple with a spaceship, and for the first time the term "Tractor Beam" is used. Smith coined it. After that, the Tractor Beam came to be used frequently in both written science fiction and, a little later, in motion pictures and television. The aliens from Metaluna use a green tractor beam to pull a small plane into their flying saucer in the 1955 movie *This Island Earth*—so far as I know, its first appearance onscreen. Tractor beams were used frequently in the television series *Star Trek* for grappling with other space vessels. It was used for the same purpose in the *Star Wars* movies, starting in 1977. In each case the space ships act like seafaring ships of an earlier era, and the tractor beams are like a higher-tech version of a hempen grappling line—it wouldn't seem believably futuristic to have a space ship towed by a mere rope.

In all these cases I am struck by one interesting detail—there is no need to explain the device at length. Normally in science fiction the author finds an excuse to explain the theory behind what is going on, often by explaining it to some innocent person connected to the story. Michel Verne does give a short explanation in pseudoscientific gibberish, but virtually no one else does. "They're pulling us up!" shouts scientist Cal Meacham as his plane is kidnapped in *This Island Earth*, and that's the only explanation given In *Star Trek* and *Star Wars* they simply call it a Tractor Beam, and let it go at that. Clearly, it was felt that no deeper explanation was needed.

This is all the more remarkable when you consider that no real physical analogue for this device exists, and (at least until recently—see the end of this piece) no one had a good theory for one. Perhaps the ready acceptance of this Tractor Beam technology lies in being able to conceive what it is doing easily—it grasps an object and pulls on it, like a rope. Analogy can easily be made with the forces exerted by an ordinary magnet. Of course, it's a poor analogy, because magnetic forces aren't very much like the tractor beam as shown. Perhaps we can draw comfort from the thought that light pressure does exist, and has been demonstrated, and that perhaps people use that analogy. As *The Encyclopedia of Science* Fiction notes:[3] "The true rationale for...the tractor beam...and the pressor beam...is that—like Faster Than Light travel—they help tell stories."

Until recently, the closest thing we had to a tractor beam is probably Optical Tweezers and Optical Levitation, as first explored by Arthur Ashkin and J. Dziedzic[4] at Bell Labs in the 1970s. They were able to levitate small dielectric spheres up to hundreds of microns in diameter with lasers and to position thin membranes. Later they learned how to use different optical laser modes, alternating light beams, and other tricks to develop the repertoire of Optical Tweezers, Optical Trapping, and Optical Cooling. Nothing reported was *quite* like a tractor beam, although forces *could* be applied along directions not parallel to the beam. When, for instance, the light was refracted by dielectric spheres, the light was refracted, and the sphere felt a sideways force as well as it having a component along the direction of propagation of the beam. Membranes placed inside an interferometer could be localized at the minima of interference patterns, which sometimes involved them being pushed *toward* the laser. But such interference patterns required a mirror to be placed *beyond* the object being localized. Other, more subtle traps required setups with crossing beams or mirrors or temporal variation to create the needed trap. Nothing was quite like the simple ray gun at a distance from the object being moved.

A measure of the success of the term "Tractor Beam," in fact, is that headline writers used it to announce developments in Optical Tweezer technology, ironically explaining an existing technology by reference to a familiar but nonexistent technology.[5] Leik Myrabo has also used the analogy for his work on Laser Propulsion, even using it directly in an article's title.[6]

All things change, and the nonexistence of Tractor Beams appears to be one of them. When this piece first appeared as an article in *Optics and Photonics News*, it was correct about tractor beams being imaginary. My reasoning for the nonexistence of such devices was that one could visualize a beam of light as a stream of photons, each carrying a quantum of linear momentum, which was directed outwards, away from the source. The light might be redirected, as when dielectric microspheres refract light passing through them. The microspheres are then directed in one direction and the light beam in another,[7] but the photons still have a net outward momentum, and only impart some sideways momentum to the microsphere. You would have to have the particle interacting with a photon in such a way that it pushed the photon forward with *more* momentum than it encountered the particle in order to have the particle pushed back toward the source of the beam, and that is a case of special pleading—it wouldn't work on just any object. It would be inelastic scattering, with the photon changing color. And it would require an energy source within the particle itself.

But there are other ways to get around the difficulties. I was surprised to find that my article was cited as a history of the "tractor beam" concept in a paper published in the journal *Optics Letters*—"On the Concept of 'Tractor Beams'" by S. Sukhov and A. Dogarin of the Center for Research in Optics and Lasers and Education (CREOL), at the University of Central Florida in Orlando.[8] As they point out, it has been suggested that one can create an effective tractor beam by using *two* collinearly propagating beams with a slight frequency shift and appropriate mode. In this way, one can create an "optical trap" without needing external mirrors. Items will be trapped at the nodes of the interfering beams. By adjusting the frequency difference, one can then cause these nodes to propagate backwards toward the sources, effectively pulling the object toward the source. My objection to Tractor Beams on the basis of conservation of momentum are answered by using two photons to produce a net, necessarily small, backwards momentum.[9] Sukhov and Dogariu go on to consider the requirements for nonconservative forces to provide a net inward momentum, all of which seem to require more than one co-propagating beam. They went on to generalize these results in a later paper,[10] "Negative Nonconservative Forces: Optical 'Tractor Beams' for Arbitrary

Objects." Here, as the title implies, they suggest methods of creating such beams for objects of arbitrary size and shape. The day of Edmond Hamilton's and "Doc" Smith's Tractor Beams may be closer than we think.[11]

Notes

1. Verne's original text, without Michel's additions, has been recently (2006) translated into English as *The Meteor Hunt* by Jules Verne, trans. Frederick Paul Walter, ed. Walter James Miller (2006). Curiously, the basic plot of the book Michel Verne released was used for the 2002 Mike Myers spoof *Austin Powers in Goldmember*. Myers must be a Verne fan.

2. Curiously, Verne had previously written a brief forward to another science-fiction novel about an inventor who travels into space in a sphere containing anti-gravity metal, and praised the author's ingenuity! This was *A Plunge into Space* by Robert Cromie, published in 1890, 11 years before Wells's book. (There had been several anti-gravity space-farers before Wells's Cavor, who was a relative latecomer to the field.) It's also unusual in that Verne was not facile in English. One line of speculation is that it was Michel who read the book and wrote the forward. It would be consistent with his greater tolerance for far-out ideas.

3. John Clute and Peter Nichols (eds.), *The Encyclopedia of Science Fiction* (1993), entry by Peter Nichols on p. 438. I'm indebted to the website http://www.technovelgy.com for some of these references, although they curiously missed the Verne use.

4. Starting with A. Ashkin, "Acceleration and Trapping of Particles by Radiation Pressure," *Phys. Rev. Lett* 24 (4): 156–159 (1970); and A. Ashkin and J. Dziedzic, "Optical Levitation by Radiation Pressure," *App. Phys. Lett.* ***18***: 283–285 (1971).

5. AP writer Brian Bergstein used it in his headline of October 2007 reporting on improvements in Optical Tweezers at MIT, http://www.livescience.com/technology/071030-ap-tractor-beam.html. It also appears in *Physical Review Focus* for T. Iida and N. Ishihara's recent work on the amplification of such light pressure forces in the presence of other scattering particles, http://focus.aps.org/story/v21/st21.

6. Leik Myrabo's article is "Propulsion Systems Integration for a 'Tractor Beam' Mercury Lightcraft Liftoff Engine," *First International Symposium on Beamed Energy Propulsion AIP Conference Proceedings* 664: 683–696 (2003), http://adsabs.harvard.edu/abs/2003AIPC..664..683M.

7. We reported on light pressure on the dielectric spheres that are oil droplets in this column before. See "The Case of the Oily Mirrors: A Locked-Room Mystery" in *Optics and Photonics News* 14 (7): 16–17 (July 2003), and chapter 31 in this collection.

8. S. Sukhov and A. Dogarin, "On the Concept of 'Tractor Beams,'" *Opt. Lett.* 35 (22): 3847–3849 (November 15, 2010).

9. The article they reference is Tomás Čižmár, Věra Kollárová, Zdeněk Bouchal, and Pavel Zemánek "Sub-Micron Particle Organization by Self-Imaging of Non-Diffracting Beams," *New Journal of Physics* 8 (3): 43 (2006). The authors have since gone on to write several developments of the concept.

10. S. Sukhov and A. Dogariu, "Negative Nonconservative Forces: Optical 'Tractor Beams' for Arbitrary Objects," *Phys. Rev. Lett.* 107: 203602 (2011).

11. A new article that covers the history of real, physical "tractor beams" is Aristide Dogariu, Sergey Sukhov, and Juan José Sáenz, "Optically-Indiced 'Negative Forces,'" *Nature Photonics* 7 (1): 24–27 (January 2013).

{ 40 }

The Rise and Fall and Rise of the Starbow

> Even the stars off to one side are showing relativistic color shifts. It's almost like a rainbow, one of those full-circle rainbows that you see on the clouds beneath you from an airplane sometimes. Only this circle is all around us. Nearest the black hole in front the stars have frequency-shifted to a dull reddish color. They go through orange and yellow and a sort of leaf green to the band nearest the back hole in black, which are bright blue shading to purple.... But the starbow itself is beautiful. It's worth the trip.
>
> —FREDERIK POHL, *THE GOLD AT THE STARBOW'S END* (1972)

The ostensible writer of these lines is a passenger aboard a starship traveling about a third the speed of light in Frederik Pohl's 1972 Hugo award-nominated novella *The Gold at the Starbow's End.*[1] Like all good writers of "hard" science fiction, Pohl had searched through the technical literature and found an intriguing scientific concept upon which to hang his story. In this case, the concept was the Doppler shifting of stars as seen when traveling at relativistic speeds. This, according to his source, would cause the stars to form a rainbow-colored series of concentric rings as a "Bullseye" target centered on one's destination—a *starbow*, as Pohl named it.

Pohl's story used the image of the starbow several times, and the title refers not to the optical phenomenon, but to the unexpected payoff of the voyage. The optical effect isn't central to the story. It made for an interesting cover to the issue of *Analog* in which the story first appeared (and to its book publication shortly thereafter). But by the time Pohl rewrote the novella into a full-scale novel in 1982, he changed the title to *Starburst*, and the description of the starbow was very much modified. Why?

In 1959, J. Terrell of Los Alamos published a paper entitled "The Invisibility of the Lorentz Contraction" in *Physical Review* in which he pointed out that the mental picture most people had of objects relativistically contracted wasn't really correct. In the following decade many others explored the implications. One who did so was Ing E. Sänger of Germany,[2] who seems to have been the first to consider the effects of relativity on an observed star field. He assumed, for simplicity's sake, that each star could be considered as a yellow monochromatic source at, say, 5,900 Å. This had the advantage of allowing him to easily calculate the color shift, from the known relativistic Doppler shift associated with the ship's velocity, **v**.

> Thus, with all Einstein numbers of flight [v/c] greater than 0.37 a major dark spot will surround the take-off star, and a minor dark spot the target star. Between the two limiting circles of these spots, all stars visible in the sky are coloured in all the hues of the rainbow, in circles concentric to the flight direction, starting in front with violet, and continuing over the blue, green, yellow and orange to red at the other end.

This is ultimately where Pohl obtained the idea, although he seems to have gotten the order of colors backwards.[3]

Paul Doherty of Oakland University in Rochester Michigan was intrigued by the image from Pohl's story and tried plotting it out (at least, that's what he says at his website). Starting from basic principles, he and John M. McKinley calculated the effects of relativistic contraction and, upon plotting the results, didn't see at all what Sänger had described. There was no completely UV-shifted "dark disc" in front nor a completely IR-shifted "dark disc" around the departure star, and the space between was not filled with the expected spectrum of stars.[4–6]

What was different between the two sets of calculations? McKinley and Doherty's Doppler-shift formula is more correct than Sänger's, but two other factors are more important. McKinley and Doherty also calculated the effect of relativistic contraction on observed intensity ("the most neglected phenomenon of special relativity," they called this). McKinley, in a separate publication,[7] called it the *Headlight Effect*, which some have changed inexplicably to *The Searchlight Effect*. They assumed real stars with blackbody-like spectra and a distribution of temperatures. This latter point was probably the most important. "Even in the forward direction," they note, "not all objects will be blue white, since an infrared object such as a protostar embedded in a dust cloud may have its Doppler-shifted temperature appropriate for orange or yellow." The very broad output spectrum of a real star also isn't as simple to treat as a single Doppler-shifting of a single wavelength. The entire spectrum must be so shifted, and the result multiplied by the frequency response of the human eye, and the color determined by color theory. Sänger's model was just too simple to provide an adequate picture of reality. "[Sänger's] model exactly reverses the role of star and eye as the 'wide band' and 'narrow band' component of the system," write McKinley and Doherty. They found that, even exceptionally close to the speed of light, no dark discs appear, and no starbow is evident. "We hope that this presentation will spell the end of the 'starbow.' At the same time we regret its demise. We have nothing so poetic to offer as its replacement, only better physics."

Pohl, learning of this,[8] commented, "What's the good of *that*?" But he modified his description of the effect when he turned the story into the novel *Starburst* two years later. He toned down the idea of the Starbow, put the colors in the correct order, and emphasized the Headlight Effect.[9] (The cover painting has the colors in the correct order, but everything else still resembles the effect as described in the earlier story, not the book). The Starbow, although a beautiful and beguiling image, disappeared. For a while.

Once an idea or image gets into the public domain, however, it never really dies. If it's not really interesting to enough people, it may fall in deference to more popular images. But if it really is engaging, or attached to a popular work, or serves a particularly useful purpose, it will be perpetuated despite its technical flaws. There is something Darwinian about the rise and fall of images.

The catalyst in this case is the rise of science fiction in popular entertainment. In particular, it's the depiction of spaceflight in popular motion pictures and television programs. One problem is that the makers of such films and programs are trying to make such an unreal and (mostly) unrealized things as spaceflight seem real to an audience that they assume is not used to such things, or to strict technical accuracy. To make the experience seem believable, these films have shown space ships "banking" in flight, or making audible sounds in space, or any of a number of unphysical things that annoy physicist and fans of "hard core" science fiction. One of the more common offenders is the *moving star field*. That's when the space ship is moving rapidly (probably relativistically) through space, and the stars appear to be moving at different rates behind the ship, depending upon how far away they are. It looks as if the ship is moving through an underwater tank filled with nearby glowing balls. The filmmakers do this because it gives a visceral and immediately-grasped sense of movement through a medium. The earliest film I've seen this effect in is *Fantasia* (1940), during the "Rite of Spring" sequence, as the camera tracks in toward the newly created earth through a cosmos of stars. It became pretty common in the 1950s, when even ships moving within the solar system frequently appeared to be passing stars (as in *Operation Moonbase*, *Flash Gordon*, and *This Island Earth*). It was ubiquitous in the TV series *Star Trek*. The *Enterprise*, of course, was supposed to be traveling at speeds greater than that of light, so who knew what it was supposed to look like? Subsequent shows and films used such a moving starfield as a matter of course, the extreme case being the elongated stars observed in the "jump to hyperspace" in 1977's *Star Wars*.

When they resurrected *Star Trek* as a motion picture in 1979 its transition to hyperluminal speed had a new fillip—the vanishing point to which it tended was now surrounded by a burst of light with a pronounced spectral separation. It probably wasn't intended to be a starbow, but it was evocative of it, and it was used in subsequent films in the series. When they launched a new television series, *Star Trek—The Next Generation*, stars observed in superluminal travel didn't merely seem to drift by the speeding starship, they were drawn out spectrally along the direction of travel. It was as if the starbow had reappeared in a new and different form—each star was not merely Doppler-shifted, but was broken down into Doppler-shifted components.[10]

At about the same time, the Starbow made a reappearance elsewhere.[11, 12] The anime series *Gunbusters* featured future astronauts traversing interstellar space by the use of future physics involving something called a "Starbow." The item was supposed to be different from Sänger's phenomenon, but references to the "Starbow" usually mention the original concept, without pointing out its flaws. At the same time, two articles appeared in the Japanese magazine *Astronomical Herald* that seems to cover both uses of the word. It's not clear that it pointed out the flaws in Sänger's theory. A Japanese website shows the same sort of relativistic starfield calculations as on Doherty's site, but in its case the Starbow *does* appear. (Not enough information is given to calculate why its image should differ from those of Doherty and others, also available on the Internet.[13–15] Both Doherty's and this site really do seem about accurate, upon inspection—at high values of **v**, approaching lightspeed, there appear to be many more blue stars near the center, and for certain directions, I suspect, you can see a sort of Starbow in that you see the trend. But it's not pronounced, nor, as Doherty's POV show, is it universal.) I still find uncritical references to the Starbow at Internet sites and in Internet fiction published as recently as 2005. Clearly, not everyone has heard the word about The Starbow's fall, or else they don't believe it. So the Starbow is likely to be with us for some time to come.

> When... Lester del Rey heard about the death of the Starbow, he sat down and wrote me up a list of umpteen ingenious ways to salvage the starbow in case I ever expand the story into a book... and this is what I really think—somebody, some time, will find out something new that will make it possible again, at least under certain circumstances.

GIVE US SOME LUCK AND WE'LL TIMELY SEND
YOUR POT OF GOLD FROM *THE STARBOW'S END*.
—FREDERIK POHL

Notes

1. Frederik Pohl, "The Gold at the Starbow's End," *Analog Science Fiction/Science Fact* **89** (1): 8–56 (March 1972). Reprinted in *The Gold at Starbow's End* (1972) and in *Platinum Pohl* (2005).

2. Ing. E. Sänger, "Some Optical and Kinematical Effects in Interstellar Astronautics," *J. British Interplantary Society* 18 (7): 273–277 (1961).

3. Pohl states in his article in *Destinies* that he got the image from an issue of *Spaceflight* shortly before he wrote his story in 1972. *Spaceflight* is also published by the British Interplanetary Society. Either Pohl mistook his source, or else an article from *Spaceflight* quoted Sänger's earlier *JBIS* article.

4. John M. McKinley and Paul Doherty, "In Search of the 'Starbow': The Appearance of the Starfield from a Relativistic Spacship," *American Journal of Physics* 47 (4): 309–316 (April 1979).

5. J. Terrell, "The Invisibility of the Lorentz Contraction," *Phys. Rev.* 116: 1041–1045 (1959).

6. V. P. Weisskopf, "The Visual Appearance of Rapidly Moving Objects," *Physics Today*, 24–27 (September 1960).

7. John M. McKinley, "Relativistic Transformation of Light Power," *American Journal of Physics* 47 (7): 602–605 (July 1979).

8. Frederik Pohl, "On Predicting the Future: Looking for the Starbow," *Destinies* 2 (1): 8–17 (February-March 1980).

9. Frederik Pohl, *Starburst* (1982).

10. *Star Trek* seems to have the color spread about right—blue toward the point of origin.

11. J. Fukue, "Visual Relativity: Starbow," *Astronomical Herald* 81 (1): 17–21 (1988) and 81 (3): 82–87.

12. Another image of the Starbow appeared in *Astronomy* 6 (4) accompanying an article by T. R. Schroeder in 1978.

13. Paul Doherty webpage, http://isaac.exploratorium.edu/~pauld/stars/PD_images_relativ.html, accessed September 22, 2012.

14. Other images of relativistic field, see http://www.madsci.org/posts/archives/Jan2003/1042727085.Ph.r.html, accessed September 22, 2012.

15. Site with relativistic images and Starbow, http://www.shekpvar.net/~dna/Publications/Cosmos/Gossary/S.html, accessed September 22, 2012.

{ 41 }

Diamonds in the Dark

Partly from its peculiar colour, partly from a superstition which represented it as feeling the influence of the deity whom it adorned, and growing and lessening in luster with the waxing and waning of the moon, it first gained the name by which it continues to be known in India to this day—the name of The Moonstone.

. . .

We set it in the sun, and then shut out the light of the room, and it shone awfully out of the depths of its own brightness, with a moony gleam, in the dark.

—WILLIAM WILKIE COLLINS, *THE MOONSTONE* (1868)

Short Round: Why they glow like that?
Indiana Jones: Legend says that when the stones are brought together, the diamonds inside them will glow

—SCREENPLAY BY WILLIAM HUYCK AND GLORIA KATZ, *INDIANA JONES AND THE TEMPLE OF DOOM* (1984)

Diamonds and gems that glow by their own internal light are ubiquitous in popular culture. They abounded in pulp literature from the 1920s to the 1940s, especially in planetary romances where they provided an exotic alternative to light bulbs. They add an extra bit of magic to motion pictures, where the Kryptonite in *Superman* glows a malevolent green, the magic gem in *Dragonslayer* glows with greater intensity and heat as its power is invoked, and the gem of V'ger in *Star Trek: The Motion Picture* glows with significance.

The use of glowing gems in visual art, such as movies, television, and comic books is hardly surprising—not only does it serve as visual shorthand for the power of the gems (perhaps only the power of exciting avarice), but it also makes them easily visible. But the idea of autonomously glowing gems, without any external source of power, goes back much farther than such modern visual images and has roots deeper than modern pop culture.[1, 2, 3]

The earliest reference I have found is in the *Histories* of Herodotus (fifth century B.C.E.), who described two columns in the temple of Hercules in Tyre, one of them made of green gemstone that shone by its own power at night. The geologist and gem expert Sydney Ball, writing in 1938, suggested that the priests of the temple might have kept a lamp in a hollow translucent green column to create the effect.

In the Indian epic *The Mahabharata*, the rajah Babhruvahana is said to have illuminated his palace with precious stones that "shone like lamps so that there was no need for any other light in the assembly."

Glowing gems began to appear regularly in classical writers from the first century B.C.E. onward. Parthenius of Nicaea, a grammaian, noted that a stone he called the Aster "flamed in the dark, hence called Ballen, the King, by the Phrygians."[4] Pliny, in his *Natural History*, writes of a gem he calls *Chrysolampis* that "is pale by day, but of a fiery nature at night." The name seems to mean "Golden Torch," and many have interpreted it as a topaz, although other stones, such as chrysolite and alexandrite have been suggested. None of these are phosphorescent, however. The natural historian Aelian tells a story about a woman who nursed an injured stork, and in return the stork brought her a gem that could light up a room "as though by many torches" in his *Historia Animalum*. The satirist Lucian of Samosota, in *de Dea Syria*, describes a statue of a Syrian goddess in the city of Hierapolis that had a gem called a *lychnis* that gives out light at night "so that the whole temple gleams brightly as by the light of myriads of candles, but in the daytime the brightness grows faint; the gem has the likeness of a bright fire." *Lychnos* is Greek for "a portable lamp," apparently the source of the name. It's not clear how serious Lucian is being—his writing ranges from what appears to be straight reporting to satiric flights of fancy. In the third century C.E., pseudo-Callisthenes (a late writer adding to the work of the real Callisthenes, a companion of Alexander the Great) wrote that Alexander found in the bowels of a fish he speared a white stone so brilliant that everyone believed it to be a lamp, and that Alexander used it as a lamp at night, having set it in a gold frame.

There are reported Chinese accounts of luminous stones, but these may be enthusiastic mistranslations.[5] Hebrew legends, recorded in the Talmud, the *Genesis Rabbha*, the *Sepher Noah*, and other sources (dating from about 400–1100 C.E.) tell of luminous stones, such as pearls that Noah used to light the Ark during its long voyage, or of Abraham lighting a city with a bowl of glowing gems.

The legends continued on into the Middle Ages. Alardis of Amsterdam, perhaps inspired by Pliny, also wrote of a *Chrysolampis* that illuminated a chapel, allowing monks to read into the night. The Carbuncle[6] at the shrine of St. Elizabeth in Marburg was also said to glow in the evenings. *The Song of Roland*, the twelfth-century French saga, tells of Saracen ships lighted by such carbuncles.[7]

Many of these reports are clearly legendary and are not intended to be real reports. Sometimes they seek to explain difficult problems (how could Noah keep the Ark lit through forty days and nights of storm?) or represent wishful thinking. But is there any reality lying at the heart of such stories?

Although knowledge of this did not reach Western society until very recently, the Uncompahgre Ute Indians used triboluminescence—in the form of mechanoluminescence—by constructing ceremonial rattles of a wooden form containing a hollow space filled with quartz crystals, all wrapped in buffalo hide. Shaking the rattles not only produced rhythmic sound but also provided intermittent flashes of light as the quartz crystals impinged on each other. The light was visible through the translucent buffalo hide.[8]

The thirteenth-century Dominican bishop and philosopher Albertus Magnus reported in *De Mineralibus et Rebus Metallicis* that some diamonds emitted light when placed in warm water. This sounds very much like another of those bits of questionable folklore, but the seventeenth-century Anglo-Irish physicist and chemist Robert Boyle was sufficiently impressed to study the diamond to see if there was any emitted light. Although he had no success initially, he eventually found that pressing on a diamond with the point of a steel bodkin would produce luminescence. Francis Bacon had observed several years earlier that sugar would produce "sparks" of light when stirred.

A century later, in 1792, Thomas Wedgwood[9] reported in the *Philosophical Transactions of the Royal Society* that he observed bright luminescence when he rubbed two diamonds together. Many others have reported on this phenomenon since. It's commonly observed during the polishing of diamonds using diamond dust, and the nature of the emission varies with the direction of polishing. The emission begins at about 400 nm before rising to a peak at 525 nm. Scratching the diamond instead of rubbing produces a different, much broader band spectrum.

In addition, many diamonds will fluoresce when exposed to ultraviolet light, glowing red or blue. But some diamonds will phosphoresce after exposure to ultraviolet light, continuing to glow faintly after the source is removed. It's conceivable that a diamond, after being exposed to the ultraviolet of sunlight and brought swiftly into the dark, might have a glow visible to dark-adapted eyes, and this might have given the ancient legends a nudge. The Hope diamond has famously displayed this behavior (it has been suggested that recording the characteristic spectrum of the glow under UV light be used to "fingerprint" diamonds), as have several other blue diamonds.[10] In one case the glow persisted for 28 seconds after exposure.

Another possibility is that some diamonds produce thermoluminescence when heated.[11]

Other crystals can exhibit very impressive triboluminescence.[12] I have seen very bright triboluminescence from alkali halide crystals containing color centers.

Of course, the effects listed here are relatively dim and of short duration—any emission of light requires a power source of some sort, and when that gives out the glow ceases. The stories of glowing gems we have are, if based on observed phenomena, amplifications of transient real effects.

The draw of glowing diamonds has led some to create their own luminous gems using artificial means. On October 4, 1921, Charles G. Smith of Cambridge,

Massachusetts, was granted US Patent No. 1,392,604 for jewelry in which "radioactive substances" were incorporated into slits cut in the jewels, allowing them to glow, much like the Radium lamps in the pulp fiction of Edgar Rice Burroughs and Robert E. Howard. Fernand Sauvagé of Paris was granted Patent No. 1,511,140 three years later for a similar idea.

The idea is still alive, although using less toxic mechanisms. Sheldon Kwiat of Great Neck, New York, was granted Patent No. 6,553,786 on April 29, 2003, for jewelry consisting of a grid of tiny diamonds on a special backing. The diamonds are selected to be a mixture of those that fluoresce under UV light and those that do not, and the backing contains tiny UV Light Emitting Diodes. When the LEDs are turned on they make the fluorescent diamonds glow and spell out a message. The example used in the patent diagram is "LOVE."

Notes

1. G. F. Kunz, *The Curious Lore of Precious Stones* (1913), 161–162.

2. S. H. Ball, "Luminous Gems, Mythical and Real," *The Scientific Monthly* 47 (6): 498–505 (1938).

3. T. Wedgwood, "Experiments and Observations on the Production of Light from Different Bodies, by Heat and by Attrition," *Phil. Trans. Royal Soc.* 82: 28 (1792).

4. Charles William King, *A Natural History of Precious Stones and of the Precious Metals* (1870), 9.

5. See Berthold Laufer, "The Diamond: A Study in Chinese and Hellenistic Folklore," *Field Museum of Natural History Publication* 184, Anthropological Series 15 (1) (1915), online at http://books.google.com/books?id=dOEqAAAAYAAJ&pg=PA56&dq=%22shining+jade%22&hl=en&sa=X&ei=MdbwT5f1GufHoQGC6-X7Ag&ved=0CEEQ6AEwAQ#v=onepage&q=%22shining%20jade%22&f=false and at http://books.google.com/books?id=ZekSAAAAYAAJ&pg=PA56&dq=%22shining+jade%22&hl=en&sa=X&ei=SefwT8beMaHaoQHpmuT6Ag&ved=0CE8Q6AEwAw#v=onepage&q=%22shining%20jade%22&f=false, accessed July 1, 2012.

6. I must admit to a sense of confusion when I hear a *carbuncle* described as a gemstone, as in the Sherlock Holmes story "The Adventure of the Blue Carbuncle." I first learned of a carbuncle as a large abcess. But it turns out that the gemstone definition is earlier, with the infirmity named after the stone as a poetic or ameliatory derivation. The title of Arthur Conan Doyle's story notwithstanding, a carbuncle is a red stone and is generally taken to be a garnet.

7. Online at http://www.jjkent.com/articles/folklore-chrysolampis-carbuncles.htm, accessed July 1, 2012.

8. http://en.wikipedia.org/wiki/Triboluminescence and references 1 and 2 therein, accessed July 1, 2012.

9. Wedgwood, "Experiments and Observations on the Production of Light from Different Bodies, by Heat and by Attrition."

10. T. M. Moses et al., "A Contribution to Understanding the Effect of Blue Fluorescence on the Appearance of Diamonds," *Gems and Gemology* **33** (4): 244–259 (Winter 1997).

11. See chapter 14, "Thomas J. Pearsall and the Ultraviolet." Pearsall was able to induce such behavior in diamonds and other substances by exposing them to the ultraviolet light of electric sparks.

12. J. R. Hird et al., "Triboluminescence from Diamond," *J. Phys. D* 40: 1464–1472 (2007).

{ 42 }

Pop Culture History of the Laser

There are several histories of the Laser[1]—for instance, Jeff Hecht's *Beam: The Race to Make the Laser* or Mario Bertolotti's *The History of the Laser* or Charles Townes's *How the Laser Happened*—concentrating on the personalities involved, or on the technical development, or some other feature. But this brief article is concerned with something else entirely—the way the Laser was perceived and understood by the nontechnical public.

The Laser is something very different from most other technical advances, most of which take many years to make a large impact on pop culture. It's often difficult to understand exactly what the breakthrough is, especially if its manifestation is something not easy to see. The public wasn't very moved by the Microwave Amplification by Stimulated Emission of Radiation, or by Magnetic Resonance. But the Laser produced, in its very first manifestation, a concentrated, directed beam of light (even if only in short bursts) that could be clearly seen. What's more, there was a ready-made analogue that had been prominent in pop culture for decades—the Ray Gun.[2] Ted Maiman's Ruby Laser, with its spiral flashtube wrapped around the exotic and expensive ruby core, even looked the part. If Javan, Bennett, and Herriott's original HeNe laser had been the first—with its weak invisible infrared beam at 1.15 microns and its resemblance to a neon sign—people would have been much less impressed.[3]

The decade of the 1950s seemed to be setting the public up for the Dawn of the Laser. "Ray Guns" may have been a staple of pulp science fiction and newspaper comics like *Flash Gordon*, *Buck Rogers*, and *Brick Bradford*, and their incarnations in radio and movie serials, but that still wasn't a very broad audience. Beginning in the 1950s, however, such futuristic ray weapons began appearing in feature films. And not only in cheap, low-budget films, but in mass-market films made in color, with high production values, and sometimes intended for a sophisticated audience. America in the 1950s was given a crash course in science-fiction weaponry by Gort's laser-like eye beam in *The Day the Earth Stood Still* (1951), to the original ray gun, H. G. Wells's Martian Heat Ray in George Pal's 1953 film of *The War of the Worlds*. The pulp-inspired epic *Forbidden Planet* gave us "blasters" that fired brilliantly colored animated beams, courtesy of Joshua Meador and A. Arnold Gillespie, and the epic *This Island Earth* featured a space war fought with bright red ray beams. *It Came from Outer Space* gave us hand-held beams in 3D, and even Disney's TV series *Disneyland* featured an episode with ray guns. By the end of the 1950s, even people who never read science-fiction books, magazines, or comic strips had been told what a ray gun ought to look like and how it ought to act.

The television show *My Favorite Martian* in the early 1960s had its hero Tim O'Hara (Bill Bixby) building a laser under the direction of the titular Martian (Ray Walston), using the ruby from his ring pumped by light from his slide projector. The basic concept is surprisingly close to correct. When he turns it on, it begins to zap objects cleanly out of existence, with no residue. So the construction is correct, but it's imagined to be a perfect Disintegrator Beam.

In the episode[4] "From Venus with Love" of the British TV series *The Avengers* astronomers are being killed off by something that looked like a ball of light and caused a high-pitched whine. Liquids in the room with the victim bubbled over. The cause turns out not to be extraterrestrial, as it seems at first, but a very terrestrial villain in a shiny-sided car who uses a laser to kill his victims. A fortuitous tape recording of the crime scene which captured the characteristic whine is heard by a scientist who hears the sound and identifies it—"That's a Laser!" At the end, the hero's bowler hat is turned from its normal black to white by exposure to the laser.

In the 1964 James Bond movie *Goldfinger* the titular villain has James Bond spread-eagled on a slab of gold, a huge laser ominously pointed down to bisect both the slab and Bond. "Do you expect me to talk, Goldfinger?" "No, Mr. Bond, I expect you to *die*!" It's an iconic moment in 1960s pop cinema. The laser itself is clearly a scaled-up version of Maiman's ruby laser, with an enormous helical flashtube. In fact, it's pretty clearly based on the publicity photo, rather than the real thing. When the laser turns on there's a burbling sound, followed by a whipcrack sound as the laser comes to life and a low whine as the continuous beam starts to cut the gold—there's no laser speckle from the beam, and it doesn't reflect or scatter from the gold as you'd expect. Instead, the gold melts, while a flame rides backward along the direction of the beam (achieved by someone holding a torch, cutting through the slab from underneath). It has a weird appearance, which is clearly the intent. Bond is not cut in half by this modern-day sawmill, of course, and the laser goes on to play an important part in Goldfinger's plot—a part it did not have in Ian Fleming's original novel (which pre-dated the laser).[5] Its presence was needed because both the heroes and the villains in the films used the latest scientific and engineering technology.

I bring these examples up because they really do show the way the laser was presented to the public by the makers of popular entertainment, and which therefore dictated how the public perceived and understood the laser. We may hope that people learn about science and technology from popular science shows, journals, and the news, but Pop culture is much more widely and repeatedly watched.

So when *Star Trek* showed lasers heating things and melting them, and knocking people unconscious ("neural impact" was the way it was described in their in-house literature), that's what people remember. Even though creator Gene Roddenbery changed the name of the device to *phaser*, because he was afraid it would turn out that lasers couldn't really do what he depicted.[6]

So in the popular mind, lasers combined real-life features (the collimated beam, the details of construction, the ability to put concentrated energy at a point) with features created to make it visually and aurally interesting (the odd flame in *Goldfinger*, the audible whine, making liquids bubble or turning black felt white) and features adapted from fictional ray guns (totally disintegrating things, stunning people, the characteristic "Zap").

Clearly better science education is called for. But the unique features of laser light were themselves often esoteric—"Coherence" was, and still is, an unfamiliar concept to the public at large. So was real monochromaticity. Photon statistics, altered side light, and mode mixing are difficult concepts to convey. Fortunately, the large coherence length of the laser lead to Leith and Upatnieks' implementation of the laser to holography. Three-Dimensional images was a concept the public could grasp, even if they didn't fully understand it.

Notes

1. Jeff Hecht, *Beam: The Race to Make the Laser* (2005); Mario Bertolotti, *The History of the Laser* (1999); Charles Townes, *How the Laser Happened* (1999).

2. See chapter 36, "Zap."

3. In fact, the publicity photo taken of Maiman behind his laser tube looked much more impressive than the actual device. They substituted a longer spiral flashtube and a longer ruby rod. See the publicity photo as figure 16.2 in Hecht, *Beam*, and compare it with the actual device in figure 15.1.

4. Season 5, broadcast January 20, 1967, written by Philip Levene, who wrote some other science fiction and fantasy based teleplays.

5. Fleming's novel actually *did* have Bond threatened with that element of Victorian-era melodrama, the circular sawmill blade. On October 7, 1973, the BBC broadcast an episode of the series *Omnibus* entitled "The British Hero," in which the scene was dramatized as it was described in Fleming's novel, circular saw and all. Christpher Cazanove played Bond.

6. Roddenbery wasn't alone in this. Science-fiction literature depicted lasers as being able to stun or paralyze, a feature clearly inherited more from fictional ray guns rather than real-life lasers. See Daniel F. Galouye's 1964 novel, *Simulacron-3*, for instance. But see also the French "light ray gun" described in *Popular Science* for January 1935, p. 17, which was said to be able to blind and incapacitate animals and people. It appears, from the description and picture, to be a flash gun with a concentrating parabolic reflector attached. http://books.google.com/books?id=lyoDAAAAMBAJ&pg=PA17&dq=%22Ray+Gun%22&hl=en&sa=X&ei=PMaNULnnK-XZoQGMw4GoCA&ved=0CFsQ6AEwCQ#v=onepage&q=%22Ray%20Gun%22&f=false, accessed October 27, 2012.

{ 43 }

Pop Culture Errors in Optics

Martin Gardner once wrote a column on "Literary Science Errors" for the magazine *Skeptical Inquirer*. There are several websites devoted to the topic, including the provocatively named *Insultingly Stupid Movie Physics*,[1] and Phil Plait's *Bad Astronomy* website.[2] There are also a few entries on it in the pop culture website TVTropes (under the headings *Ranged Energy Attack Tropes* and *Frickin' Laser Beams*[3]). But very little of it is centered on errors in optics. I'd like to remedy that.

The one notable optical error pointed out by Gardner[4] was in William Golding's *Lord of the Flies* in chapter 2, where they start a fire using the lenses of the eyeglasses belonging to the myopic Piggy. The problem is, of course, that lenses that correct for myopia are negative and will not concentrate light as a positive magnifying glass will. Golding was trying, as any good novelist, to say several things with his choices—the glasses also function as a symbol of civilization, and Piggy might be seen as myopic in other ways besides physically, but it could have worked as well with a hypertropic character. In any event, James K. Huhn pointed out in 1991 that most negative lenses are meniscus lenses, and that if one filled the concave interior with clear water the combination would act as a net positive lens, so one could, in fact, use Piggy's glasses to start a fire.[5]

A very common source of error is the depiction of lasers and other beam weapons. These fall into three main classes. The first is that beams move very slowly, so that you can see it striking out. This is probably an "intentional" error, done because the film makers want to be sure the audience realizes that the laser is a weapon, and that it's proceeding from one place to another. They don't try to show bullets going from one place to another slowly, of course. Light, being much faster than bullets, ought to be as apparently instantaneous, but film makers tend to err on the side of caution. So Gort's ray-blasts in the original *Day the Earth Stood Still* can be seen to take time to move from Gort to the things he melts, and the "blasters" in *Forbidden Planet* are pulsed beams whose interruptions take an incredibly long time to travel to their targets. The "phasers" in the original episodes of TV's *Star Trek* moved slowly enough that you could perceive them starting out. The ultimate expression of his was the 1969 episode "Wink of an Eye," where beings living in a faster time frame are able to casually walk out of the path of a beam, which they see moving slowly toward them. It's hard to cut them slack for an error that egregious. The trope continued, however. In the original *Star Wars*, the beam weapons are still moving slowly enough to see them do so.

The second error is the visibility of lasers and other beam weapons from the side. Again, this is probably intentional, not only to clue the audience in onto who is shooting what, and because it looks very pretty to see colored beams going from place to place. But, again, bullets from contemporary guns aren't visible as they fly from weapon to target, so this convention isn't necessary.

Sometimes the film makers justify this by using an aerosol scattering spray to make the beams visible, as when a jewel thief breaks into a museum and sprays something to make the detector beams evident. Of course, the sort of criss-crossing laser beams restricting access to the object being protected is unrealistic—motion sensors and infrared sensors would be more effective, but would deprive them of the opportunity to have the thief athletically dodge those beams. Once again, style and image trump accuracy.

The third error is to not depict lasers realistically. Even though lasers have become a literally everyday item, with lasers in DVD players, supermarket scanners, and laser pointers, I have still to see a fictional TV or movie laser that produces laser speckle. When lasers were still new things, they could be depicted as immensely powerful beams, with even small ones capable of cutting through metal. Lasers are still often shown as too powerful. And they never seem to need air or water cooling. Also, lasers tend to be much noisier in film than in real life, and the noise isn't the right sort. The beams themselves make high frequency trills, but you never hear the sound of fans or of water pumps.

The 1972 James Bond film *Diamonds Are Forever* added another error by having a character state that "the first laser beam was generated with a diamond" (thus explaining the supervillains quest for the titular diamonds, in this case to build a laser weapon), thus confusing the ruby Ted Maiman used in his original laser with the sexier diamonds. In fact, no lasers were built using diamonds as a host until the early 1990s, when Rand and DeShazer constructed one using color centers in diamonds.[6]

Mention of *Star Wars* recalls the "Light Sabers" of that franchise. Nothing quite like them existed before in science fiction, although the "force (field) knives" used in some 1950s fiction, or Larry Niven's "Variable Swords" give the impression that they'd look similar to those Light Sabers. Those devices weren't supposed to be optical instruments, but then again, the *Star Wars* franchise in its movies and television shows doesn't really say exactly what a Light Saber is. In any event, *Star Wars* is essentially juvenile fantasy writ large, and there's not much point in criticizing its use of what is simply a fantastic sword.

Returning to more realistic works, the thrillers written by Frederick Forsyth are much more realistic and well-researched than the general run, but even he misrepresents Image Intensifiers in *The Devil's Alternative*. They are described as devices that concentrate the existing light, rather being active devices that illuminate a target with infrared light. This makes them sound rather like lenses. There's no mention of the microchannel amplifiers that allow them to make the most of dim light, and which would have been a tempting subject for Forsyth's usually excellent descriptions.[7]

Another usually reliable source that stumbled is TV producer and host James Burke, best known for his "Connections" series on PBS and on The Learning Channel. For the second episode of his PBS series *The Day the Universe Changed*, "Light from Above," he describes the work of Theodoric of Freibourg, who described the path of light through spherical raindrops that created primary and secondary rainbows. His work is rarely mentioned, and Burke does a great service in bringing him up. But Burke's re-creation of Theodoric's experiments is simply wrong, and doesn't look remotely like the real experiment.[8] A similar experiment is described and illustrated in Minnaert's classic book *The Nature of Light and Color in the Open Air*,[9] and I have performed it in lectures on rainbows.

Finally, there are comic books. Perhaps I shouldn't point out their errors, since these have classically been low-budget, quickly turned-out pieces of juvenile entertainment, and so ought to be cut more slack even than the *Star Wars* films. But they also can play a role in education by stealth (much as I hope these little essays to be). So when Superman, in a 1960s story, builds a chain of giant lenses to form a complete circle, and projects his heat vision through them, hoping to increase the output enormously, I can't help but lament a teaching moment lost. His endless chain of lenses is really the same as a commonly used model of a laser resonator—but it lacks an essential ingredient—a gain medium. Superman's heat vision ought to be attenuated by each passage through a lens, rather than amplified.

Similarly, The Atom's encounter with the provocatively named Dr. Light in *The Atom* No. 8 in 1963 has the size-changing superhero avoiding Dr. Light's weapon, which transfers momentum with light (a clever idea there), by shrinking down so small that he can avoid the individual photons. These are depicted as perfect, monocolored spheres.[10] When I first read it, I did not think about the fact that I could no more see an individual photon not directed toward my eye than I could see a laser beam in clear medium. The Atom, after all, would have to be seeing photons given off by the photons themselves. At least the story taught me about photons and what they were—but it was by contemplating what a beam of light would look like to someone in an unusual state that Einstein first conceived of Special Relativity.

I suppose that, on balance, I shouldn't complain about that lost opportunity. But consider another such comic-book moment that succeeded in exploiting that Teachable Moment. In *Green Lantern* No. 8, the superhero observes something being shrunken down to minute size, but the capabilities of his Power Ring (the magical device that is the source of his power) are exceeded by a surprisingly real physical limitation.[11] He cannot see the object once it has shrunken below the wavelength of light. That was a properly exploited Teachable Moment. The idea that the wavelength of light itself served as a limitation on resolution came as a revelation to me, but it made sense as I contemplated it. It made me think about the ramifications—you could improve resolution by using shorter wavelengths, for instance. You can't ask more from a Teachable Moment than that.

Notes

1. http://www.intuitor.com/moviephysics/, accessed September 30, 2012.

2. http://www.badastronomy.com/index.html, accessed September 30, 2012.

3. http://tvtropes.org/pmwiki/pmwiki.php/Main/FrickinLaserBeams, accessed September 30, 2012.

4. Martin Gardner, "Notes of a Fringe Watcher: Literary Science Blunders," *Sketical Inquirer* 19 (1): 14ff. (January/February 1995); reprinted in *Weird Water and Fuzzy Logic* (1996), 117–130.

5. James K. Huhn, "You can be Myopic and Still Survive on a Desert Island," *The Physics Teacher* 29 (9): 577 (December 1991).

6. Steven C. Rand and Larry G. Deshazer, "Visible Color Center Laser in Diamond," *Optics Letters* 10 (10): 481–483 (October 1985).

7. Frederick Forsyth, *The Devil's Alternative* (1979).

8. James Burke, *The Day the Universe Changed*, Episode 2, "In Light of the Above," broadcast on PBS (USA), March 26, 1985. The substance of the episode, although not the improper replication of Theodoric's experiment, appears as chapter 2 of the companion volume *The Day the Universe Changed* (1985), 19–54.

9. M. Minnaert, *The Nature of Light and Color in the Open Air* (1954).

10. "Lockup in the Lethal Light Bulb," *The Atom*, No. 8 (September 1963). Script by Gardner F. Fox.

11. "The Challenge from 5100 A.D.," *Green Lantern*, No. 8 (September–October 1960), John Broome, script.

{44}

Pop Spectrum

One of the questions that must greet anyone working in spectroscopy is the one addressed by spouses, children, relatives, and friends not in the sciences: "What do you *do* in there all day?" And so all of the professors, students, and technicians have to try to explain in layman's terms just what spectroscopy is, and why they do it. Not everyone has the advantage of a friend or relative working at a spec lab, however. What do *they* think is going on? What is the pop culture perception of the spectrum and spectroscopy?

I'm not going to consider the rainbow, or interference bands in soap bubbles and oil films. How is the practice and art of splitting light into its constituent colors and interpreting their intensities perceived by the general public?

Pretty vaguely, for the most part. Spectroscopy is one of those tools seen in forensic science television series such as *CSI*, one of a battery of tests used in analyzing unknown substances. But the instruments are known, if at all, by their casings, in the current lab equipment style, and to the general public it doesn't matter whether the instrument is a Fourier Transform Infrared Spectrometer or an Atomic Absorption Spectrometer or a Mass Spectrometer. It's a magical device that spits out correct and infallible answers in no time at all and without that troubling ambiguity. I know that I would have loved to have had one of those when I found a very regular absorption structure in one of my superoxide-doped laser crystals that was turning the normally clear crystals brown. It took me weeks to track down the source of that alien absorption. I finally tumbled to the fact that it was due to ozonide only by analogy with the similar unexpected absorbing species S_3^- seen in sulfur-doped crystals. I sure could have used a computer-assisted analysis routine with a built-in reference library.

Usually one only hears the results, but in the film *Outland* you actually see a spectrum with peaks. This the doctor immediately interprets as a particular drug on sight, without having to consult any references. But I'll let that slide in the interests of compact and concise story-telling. At least you saw an analytical mind at work there. Even in most science fiction, ironically, you generally don't see the mechanics of analysis—you get instant results.

There's very little of the physical nature of light or the spectrum in most pop literature. There are a very few exceptions, but they're from the end of the nineteenth and the beginning of the twentieth centuries. In 1893 Ambrose Bierce published the short story *The Damned Thing*,[1] about an invisible creature that attacks people. The creature is invisible because, as Bierce's narrator tells us,

> At each end of the solar spectrum the chemist can detect the presence of what are known as "actinic" rays. They represent colours—integral colours in the composition of light—which we are unable to discern. The human eye is an imperfect instrument; its range is but a few octaves of the real "chromatic scale." I am not mad; there are colours that we cannot see.
>
> And, God help me! The Damned Thing is of such a colour!

It's an explanation that makes sense, as long as you don't think about it too much. Whatever has such a characteristic infrared or ultraviolet "color" still has to interact in some fashion with visible light.

Edgar Rice Burroughs's novel *A Princess of Mars* is science fantasy, rather than science fiction, a work having the trappings of science fiction but not aspiring to anything resembling scientific accuracy or technical rigor. So we shouldn't blame him for his errors in science, but rather credit him for knowing what he does and using it as a plausible-sounding rationale for his background.

In Burroughs's universe, there are at least two rays beyond the visible spectrum of seven colors, and these "eighth" and "ninth" rays have wonderful properties.[2] The ninth ray is used to manufacture oxygen in the Martian atmosphere factory.

> This ray is separated from the other rays of the sun by means of finely adjusted instruments placed upon the roof of the huge building, three-quarters of which is used for reservoirs in which the ninth ray is stored.

The eighth ray is also remarkable:

> This ray, like the ninth ray, is unknown on Earth, but the Martians have discovered that it is an inherent property of all light no matter from what source it emanates. They have learned that it is the solar eighth ray which propels the light of the sun to the various planets, and that it is the individual eighth ray of each planet which "reflects," or propels the light thus obtained out into space once more.

Burroughs doesn't say where these rays lie relative to the visible spectrum (indeed, he suggests that they are themselves visible, just unknown to terrestrial vision and science). One naturally suspects they are infrared, but the atmosphere-making properties suggest the high energy photons of the ultraviolet. If one were feeling generous, one could argue that UV rays could provide both photochemical effects and some form of laser propulsion, but that's more thought than Burroughs probably put into it.

Probably the most extravagant use of the spectrum in pop literature is in H. P. Lovecraft's 1927 story "The Colour out of Space."[3] Lovecraft was an extremely creative and influential author, whose works are still in print and avidly read. In "The Colour out of Space," he managed to make a color the monster in his story. It all begins with the fall of a strange meteorite at the Gardner farm in the fictional town of Arkham, Massachusetts. Professors from nearby Miskatonic University examine it and take samples, which are subjected to a battery of tests. Lovecraft was an amateur astronomer and science fan, and his descriptions of these tests are pretty accurate.

> and when upon heating before the spectroscope it displayed shining bands unlike any known colours of the normal spectrum there was much breathless talk new elements, bizarre optical properties, and other things which puzzled men of science are wont to say when faced by the unknown.... They had uncovered what seemed to be the side of a large coloured globule embedded in the substance. The colour, which resembled some of the bands in the meteor's strange spectrum, was almost impossible to describe; and it was only by analogy that they called it colour at all.

Eventually, the meteorite dissolves and the evil color seems to infest all things growing on the farm, poisoning them. In fact, Lovecraft's description of the failing health of the crops, animals, and people living on the farm strongly suggests to the modern mind contamination by some chemical substance:

> Specimens could be analyzed.... under the spectroscope both samples gave off an unknown spectrum, in which many of the baffling bands were precisely like those which the strange meteor had yielded in the previous year.

But in Lovecraft's universe, it was the color itself that caused the blight, and which had a sort of life of its own. "It was just a colour out of space—a frightful messenger from unformed realms of infinity beyond all Nature as we know it."

Many years later, Lovecraft's story was turned into a feature film starring Boris Karloff, but the aspect of a monstrous color was completely dropped. The title was changed to the more lurid and salable *Die, Monster, Die*. The meteorite produces its effects by a physical contamination, probably radioactive in nature. There is no spectral analysis in the film, but at the end Karloff's character does glow with a weird greenish light. (When pressed to actually show a weird or unusual color, the default is invariably greenish.) A 1987 adaptation of the story bore the title *The Curse*, and was even less memorable.

It might seem odd that appearances of spectroscopy and odd colors in popular literature are confined to the first quarter or so of the twentieth century, and virtually ignored or taken for granted since, but it's actually pretty typical. When a new technology or phenomenon appears, it becomes the item of interest in popular stories. When television first became common, there were pulp magazine stories of evil television sets (I kid you not). After the first atomic tests, radiation became the source of B-movie monsters and other wonders (Spiderman originally resulted from a student bitten by a radioactive spider). As new effects and technologies come into the limelight, they take over this role. (In the recent movies, Spiderman was the result of a student being bitten by a genetically engineered spider. Radiation is passé.) Spectroscopy was already an established discipline by the time Bierce wrote, but the fledgling field of science fiction and science fantasy literature hadn't yet had a chance to embrace it. Over the next few years Bierce, Burroughs, Lovecraft, and a few others used the still unfamiliar discipline to give an air of verisimilitude to their fantasies. But now the technology is old and established. There was a flurry of interest in lasers when they were new, but it focused on the potential destructive power of these coherent beams. Laser spectroscopy was too esoteric to grab the public interest and couldn't really compete with real-life Death Beams.

Notes

1. Ambrose Beirce, *The Damned Thing*, first published December 7, 1893, in *Tales from New York, Town Topics*. It's online at Project Gutenberg at http://www.gutenberg.org/ebooks/23172, accessed September 30, 2012.

2. Edgar Rice Burroughs, *A Princess of Mars* is online at Project Gutenberg as well: http://www.cs.cmu.edu/~rgs/pmars-table.html and http://www.gutenberg.org/ebooks/62, accessed September 30, 2012. There have been two recent cinematic adaptations of the novel, neither of which uses this "fact." The 1988 film *Total Recall* steals this bit of Martian fiction from Burroughs for its climax, but substitutes an imaginary mineral/element/ore, *Terbinium* for oxygen generation. The 2012 remake abandons Mars and oxygen-making altogether.

3. Lovecraft's story was first published in *Amazing Stories* 2 (6): 557–567 (September 1927). It is available at several online locations, including http://www.yankeeclassic.com/miskatonic/library/stacks/literature/lovecraft/novellas/colouro.htm, accessed September 30, 2012.

{ 45 }

The Telephote

In the American magazine *The Forum*, February 1889, there appeared an interesting piece of fiction. Although attributed to Jules Verne, the piece is now believed to be a collaboration between Verne and his son Michel, and exactly who wrote what is still being debated. This was the first appearance of the piece, and it was surprisingly in English, not as a translation of an originally French article. "In the Year 2889" was nominally the work of a future reporter, which provided an opportunity for the prediction of scientific and technological wonders to come. Among these was a broadcast from the moon, including both sound and visual transmission.[1]

> The telephote! Here is another of the great triumphs of science in our time. The transmission of speech is an old story; the transmission of images by means of sensitive mirrors connected by wires is a thing but of yesterday.

Verne must have thought his extrapolation of such a device precisely a thousand years into the future was too timid, because four years later he included the *telephote* among the technological devices used in his story "Le Chateau des Carpathes" ("The Carpathian Castle"), set in contemporary times. In this story, one of the first to feature the faking of supernatural occurrences by using scientific tricks, his villain spies on his victims using what are essentially microphones and television cameras. (This novel is also the first fictional work associating vampires with Transylvania, three years before Bram Stoker would publish *Dracula*).

When I first encountered this story, I thought that Verne came up with his proto-television and the name for it by simply extrapolating from the telephone,[2] but the real story is more complex and interesting. Verne was a tireless reader of scientific magazines, and there was much there to draw from regarding the transmission of images in the technical literature of his day.

The word *telephote* appeared on the scene in 1880. Alexander Graham Bell, inventor of the telephone, mysteriously sent a sealed package to the Smithsonian Institution. The contents were to be revealed at an upcoming meeting of the American Association for the Advancement of Science.[3] The *Physical Notes* column in the April 15 issue of *Nature* noted that there were rumors that the new device was to be called the *telephote* or *diaphote*, and that this would be able to "transmit light as the telephone transmits sound." It went on to say that two American inventors laid claim to such a device.[4]

Devices for the transmission of pictures over telephone lines had existed for years, pre-dating the invention the telephone. Such devices took a while to transmit such pictures, however (as does a current-day FAX machine, for that matter), and the excitement and the wording of the *Physical Notes* item clearly suggests something faster and more direct that time-consuming transmission of a printed image. "If this instrument should prove to be as effective as the telephone," said the *English Mechanic and World of Science* on July 23, 1880, "It will lead to results which will startle even this advanced age. Surely the men of the nineteenth century will wonder if a merchant, sitting in his office in London, should be able, by means of the telephote, to examine a sample of merchandise in Glasgow, and, by means of the telephone, to strike a bargain." Similar stories and reprints appeared in many journals of the day.[5]

The 1881 edition of *Annual Register of World Events*, reporting the previous year's items, noted with disappointment that "when, however, Professor Bell actually made known the instrument he had invented, it proved to be one for the transmission of sound by the agency of light. The *photophone*, as his instrument is called, solves this problem in a way which is almost startling in its simplicity." It describes how Bell's photophone uses a modulated light source to send light signals to a selenium cell placed at the focus of a parabolic mirror, causing a modulated signal in his receiver.[6] A plaque at the corner of 13th street and K street in Washington, D.C., marks the spot where he demonstrated his invention. Today it is seen as the ancestor of fiberoptic telecommunications.

But the seed of the idea of instantaneous image transmission had been planted and given a name. This was not the first time such an idea had surfaced—even before rumors of Bell's device started floating, cartoonist George duMaurier had published a cartoon in *Punch's* Almanac in December of 1879, depicting "Edison's Telephonoscope," and showing a British couple conversing with their tennis-playing daughter in Ceylon as they watched her image on a very wide-screen television-like device (fig. 45.1).[7]

Three years later Albert Robida, a French author and artist, published *Le Vingtième Siècle* ("The Twentieth Century"), a novel about life in the next century, complete with illustrations he himself produced. One of the marvels of the next century was the *telephonoscope*. To judge from the name, he was inspired by DuMaurier, but Robida's device had an elliptical screen, and he showed it in far more applications—as a teaching device and an entertainment device as well as a communications item. He followed up this book with a pamphlet on the telephonoscope and two more books on the future.[8]

Within a couple of years, workers were claiming success, or at least partial success, in producing such an instrument. Their devices used, like Bell's, light-sensitive selenium cells to detect light levels and turn this into an electrical signal that could be transmitted over wires. Some works literally took it no farther than this, leading

FIGURE 45.1 *"Illustration of the Edison Telephonoscope" by George du Maurier, from* Punch's Almanac, *December 1879.*

to the following statement from W. de W. Abney, which appeared in a number of journals in 1880[9]:

> Regarding the *diaphote*, *telephote*, or whatever you like to call it, I am very despondent. I believe that this kind of photography is an impossibility, for the reason that one wire can never carry an infinite number of electrical impulses, which, on reconversion into chemical energy, shall give various gradations of light and shade simultaneously on a surface.... A multiplicity of conducting wires is evidently impossible if accuracy be required.... I again assert that—as far as we can see by the eye of science at present—there is no probability of an image reflected in a mirror at one end of a wire being reproduced at the other end at one operation. Anything short of this would not be true telephotography.

In 1880 John Perry and William Edward Ayrton of the London Technical College suggested how a *telephote* might be made to work, using just such a system segmented by what we would today call pixels, and with signals running along individual wires as Abbey despondently suggested. Inspired in part by DuMaurier's drawing, they proposed using galvanometer-like shutters. Their article, "Seeing by Electricity,"[10] appeared in *Nature* and was reprinted elsewhere. It was also described in Alfred Ritter von Urbanitzky's book *Electricity in the Service of Man* (for which Perry wrote the introduction).[11] Others, such as Denis Redmond, G. R. Carey, William Edward Sawyer, Sheldon Bidwell, and others put in claims at about this time.

One who patented his work, but did not publicize or exploit it at the time, was a young student, Paul Gottlieb Nipkow, who saw that one way to scan across an image and select different "pixels" to encode was to use a rapidly spinning disc perforated with a spiral series of holes, so that a single detector could sense the intensity of light at that portion of the field and send the signal to a corresponding light source in the receiver. Nipkow's Disc was to become the mainstay of electro-mechanical television sets until the 1920s.[12]

An article in the *Electrical Engineer* in 1889 sums up reactions to these many claims[13]—people recalled what had happened with Bell's supposed telephote, and weren't going to be fooled again.

> We are now hearing a good deal about the "telephote," "telectroscope," etc.... and we have been treated to affecting stories, which movingly recite that Edison is going to do this; while a certain French savant says that he has already done it—almost, and that an interview has been arranged, where each is to spread before the other a true account of what has been done, so that there shall be a fair division of labor, and glory.... But, to parody Patrick Henry: —I care not what others may do, so long as they do not invent a plan for "smelling by electricity."

So by 1889 there were plenty of suggestions for such schemes, and the name "telephote" wasn't a Vernean neologism, but was much in the air. Why is it, then, that we don't use the term "telephote" anymore?

The word "television" was apparently first used by the Russian scientist Constantin Perskyi on August 25, 1900, at the First International Congress on Electricity, an event held at the 1900 Exposition Universelle in Paris. It has been suggested that his use of the word was a translation to French from Russian *televidenie*, which in turn was derived from the German word, *Fernsehen*. If so, it was an interesting choice. "telephote" derives from two Greek roots for "far" and "light," while "television" is one of those hybrid words hated by purists, with both Greek and Latin roots. It derives from Greek "tele" and Latin "visio," to produce "far vision," which is arguably more descriptive of the device.

It's important to note that there was a distinction—"television" was associated with the entire art and process of seeing instantaneously over distance, while "telephote" referred to the device that was used for this. It's analogous to the way we use a device called a *camera* for the process called *photography*. Early usage supports this. Hugo Gernsback, publisher of popular magazines on electricity, wrote an article entitled "Television and Telephot" in the December 1909 issue of *Modern Electrics*, consistently using *telephot* for the device and *television* for the process. He seems to have used it consistently this way throughout his career—in 1928 Gernsback started transmitting television signals (using a mechanical television system based on Nipkow's disk) that his readers were supposed to pick up by building their own sets from designs in his magazines. He also included in his magazines fictional stories of the future that he termed "scientifiction." He himself wrote and serialized the adventures of "Ralph 124C41+," beginning in the April 1911 issue of *Modern Electrics*. In that very first episode it describes Ralph communicating with the wondrous Telephot. (Gernsback has been hailed by some as the "Father of Science Fiction." The awards given by the World Science Fiction Society are named "Hugos" in his honor.)

Other people used the term as well. An 1894 book, *The Dweller on Two Planets, or The Dividing of the Way* by Frederick Spencer Oliver (which reads like science fiction, except that Oliver apparently believed he was "channeling" the thoughts of Phylos the Thibetan, a real person from ancient Atlantis) describes the Atlanteans communicating with the *Naim*, a "combined telephone and telephote." Arthur Benjamin Reed's 1916 mystery novel *The Social Gangster* has a "Franconi Telephote" as the McGuffin driving the story. As recently as 1935 the term was still appearing in engineering texts and occasional literary works.

But its time was past. A French telephoto lens design popular with photographers in the 1920s was called the *Telephot*,[14] and we continue to call a long-distance optic a "telephoto lens." This ambiguity probably hastened the term's demise as an electronics word. "Television" was more descriptive and more intuitively grasped, and people began referring to "television apparatus" or "television sets," instead, and by 1950 this was frequently shortened to simply "television"

Notes

1. The original story, which may be wholly the work of Michel Verne, was written in October-November 1888 and appeared in *The Forum* 6: 662–677 (February 1889). The French version, rewritten at least in part by Jules Verne, appeared as "La Journée d'un journaliste américain en 2890," *Journal d'Amiens, Moniteur de la Somme*, January 21, 1891 (see http://www.jv.gilead.org.il/biblio/stories.html, accessed October 1, 2012), although another site lists its publication in *Le Petit Journal, supplement illustré* for August 29, 1891 (http://www.jules-verne.ca/vernebooks/jvbkjournal2890.html, accessed October 1, 2012). An English translation of the French text appeared in the Arco/Fitzroy volume *Yesterday and Tomorrow*, trans. and ed. by I. O. Evans (1965) and in *The Jules Verne Companion*, ed. Peter Haining (1978), which includes an illustration from the original American publication. The original English edition is available online at http://www.readbookonline.net/readOnLine/17614/ and here http://manybooks.net/titles/vernejul1936219362-8.html, among others, accessed October 1, 2012. The quotation here is from Evans's translation.

2. Editor and translator I. O. Evans evidently thought the same thing, judging from his comments preceding the story on p. 141 of the Ace paperback edition.

3. For this, and much of the early history of Television, see George Shiers assisted by May Shiers, *Early Television: A Bibliographic Guide to 1940*, Garland Reference Library of Social Science, Vol. 582 (1997). The relevant reference is on p. 18.

4. *Nature* 21: 576 (1880).

5. *The English Mechanic and World of Science*, No. 800: 484 (July 23, 1880).

6. *Annual Register of World Events* 122: 447 (1881).

7. Reproductions of the picture appear on the Internet here http://www.terramedia.co.uk/Chronomedia/years/Edison_Telephonoscope.htm and here http://en.wikipedia.org/wiki/

Telephonoscope, among other places, accessed October 1, 2012. One has to ask—Why Edison rather than Bell as the inventor of a communication device? This was only three years after Bell had patented the telephone, and over a decade before Edison's work on the motion picture. He certainly invented several electrical devices, but his most popular invention before this was the phonograph. Perhaps the date is significant—December 1879 was when Edison unveiled his incandescent lamp.

8. See, for example, R. W. Burns, *Television: An International History of the Formative Years*, History of Technology Series, Vol. 22 (1998), 79.

9. *British Journal of Photography*, June 11, 1880, p. 283; also in *Journal of the Society of the Arts* 28: 615 (May 28, 1880).

10. *Nature* 21 (547): 589 (1880).

11. Alfred Ritter von Urbanitzky, *Electricity in the Service of Man: A Popular and Practical Treatise on the Applications of Electricity in Modern Life*, trans.from the German ("with copious additions") by R. Wormell (1886), 750–752.

12. See Shiers and Shiers, *Early Television*, 22, who quote extensively from an interview of Nipkow in *The New York Times*, August 6, 1933.

13. *Electrical Engineer* 8: 423 (October 1889).

14. See, for example, Chester F. Stiles, "Telephotography," *American Photography* 9 (7): 388-394 and it is available online at: http://books.google.com/books?id=8oVRVHxnDQ8C&pg=PA388&dq=%22American+Photography%22+telephoto&hl=en&sa=X&ei=6TR_UeX4O4rqoAHYh4GABg&ved=0CDsQ6AEwAA#v=onepage&q=%22American%20Photography%22%20telephoto&f=false, accessed April 29 2013. Stiles goes on to observe that the term *Telephot* was also used by Georges Rignoux at the same time to describe yet another method of sending images via telegraph.

{AFTERWORD}

There you have the best of a decade or so, my essays on the odd and interesting in Optics, grouped around the general topics of Historical, Weird Science, and Popular Culture, with quite a bit of overlap between those groupings. From this I think the reader may learn more of my interests than of Optics. But there is a method in my madness.

History of a discipline, as I stated in the Introduction, is rarely covered very well in any field, and there are always interesting nuggets to be found by exploring the history of ideas, their development, and especially the missteps on the way to our understanding.

Weird Science is often my asking the odd question—What about those rumors of edible lasers? Are there really any, and if not, why not?—that leads you into unexpected paths. All of those lasers that have been discovered and written up—How many are there? And why are so few in use?

Pop Science is the one that's really dear to my heart. My own interest in science was stoked by museums, presentations at World's Fairs and the like, television shows, and popular books and articles on science. It's not only interesting of itself, but it's a way to interest people about what science is up to and to impress future scientists. I try to relate optics to common experience, and to draw lessons from the things people see and how they're treated in popular culture. There's a divide between the way subjects are taught and the questions people naturally ask. When people want to know about magnetism, for instance, they really want to know about the ferromagnetism they encounter and feel everyday, but both textbooks and popular treatments give them Maxwell's equations and the like, and say that ferromagnetism is too complex a topic. When people ask about the rainbow, they are usually told about raindrops acting like prisms to separate light, but that obscures the actual mechanism and completely ignores the interference aspect of the rainbow. I think that if you're going to write about science, you ought to address the things that people really do encounter, and at least give them some insight into the way they work, rather than ignoring the part that's not so easy to explain, or simply saying it's more difficult. To give an example of a popular question from the world of optics, what should the optics popularizer say when the layman asks where on a color chart "brown" resides? There's no *Brown* on a CIE chromaticity diagram, after all. So is it a color, or not? If it isn't, why can my color monitor give a great reproduction of it using only the standard phosphors? At which point I'll seemingly violate my own rule and go no further, except to provide the suggestion for the reader to follow up that the color diagram is itself not a full representation of all the aspects of a color, but a normalized two-dimensional representation.

In any case, this is what I have for now. Nothing goes on forever. As one of my professors remarked, in order for a laser to truly have a single frequency, it would not only have to have been running since the beginning of time, with no changes in intensity, it would also have to be left on for all eternity. Even forgetting about physical line-broadening mechanisms, the mere fact that the laser is turned off means that the light cannot be truly monochromatic. A really single-frequency laser would have to be left on perpetually.

To which another professor quipped, "Not in *my* lab!"

—END—

{ INDEX }